SpringerBriefs in Applied Sciences and Technology

Nonlinear Circuits

Series editors

Luigi Fortuna, Catania, Italy
Guanrong Chen, Kowloon, Hong Kong SAR, P.R. China

SpringerBriefs in Nonlinear Circuits promotes and expedites the dissemination of substantive new research results, state-of-the-art subject reviews and tutorial overviews in nonlinear circuits theory, design, and implementation with particular emphasis on innovative applications and devices. The subject focus is on nonlinear technology and nonlinear electronics engineering. These concise summaries of 50–125 pages will include cutting-edge research, analytical methods, advanced modelling techniques and practical applications. Coverage will extend to all theoretical and applied aspects of the field, including traditional nonlinear electronic circuit dynamics from modelling and design to their implementation.

Topics include but are not limited to:

- nonlinear electronic circuits dynamics;
- Oscillators;
- cellular nonlinear networks;
- arrays of nonlinear circuits;
- chaotic circuits;
- system bifurcation;
- chaos control;
- active use of chaos;
- nonlinear electronic devices;
- memristors;
- circuit for nonlinear signal processing;
- wave generation and shaping;
- nonlinear actuators;
- nonlinear sensors;
- power electronic circuits;
- nonlinear circuits in motion control;
- nonlinear active vibrations;
- educational experiences in nonlinear circuits;
- nonlinear materials for nonlinear circuits; and
- nonlinear electronic instrumentation.

Contributions to the series can be made by submitting a proposal to the responsible Springer contact, Oliver Jackson (oliver.jackson@springer.com) or one of the Academic Series Editors, Professor Luigi Fortuna (luigi.fortuna@dieei.unict.it) and Professor Guanrong Chen (eegchen@cityu.edu.hk).

More information about this series at http://www.springer.com/series/15574

Luca Patanè · Roland Strauss
Paolo Arena

Nonlinear Circuits and Systems for Neuro-inspired Robot Control

 Springer

Luca Patanè
Dipartimento di Ingegneria Elettrica
  Elettronica e dei Sistemi
University of Catania
Catania
Italy

Paolo Arena
Dipartimento di Ingegneria Elettrica
  Elettronica e dei Sistemi
University of Catania
Catania
Italy

Roland Strauss
Institut für Entwicklungsbiologie und
  Neurobiologie
Johannes Gutenberg Universität Mainz
Mainz
Germany

ISSN 2191-530X   ISSN 2191-5318   (electronic)
SpringerBriefs in Applied Sciences and Technology
ISSN 2520-1433   ISSN 2520-1441   (electronic)
SpringerBriefs in Nonlinear Circuits
ISBN 978-3-319-73346-3   ISBN 978-3-319-73347-0   (eBook)
https://doi.org/10.1007/978-3-319-73347-0

Library of Congress Control Number: 2017964257

Printed on acid-free paper

This Springer imprint is published by Springer Nature
The registered company is Springer International Publishing AG
The registered company address is: Gewerbestrasse 11, 6330 Cham, Switzerland

Preface

The brain is the ultimate challenge when it comes to understanding the basis of what makes living beings able to survive in hostile environments, adapt their lifestyle to changing environmental conditions or adapt to body damages by recruiting new neural circuits to regain limb control. Unraveling the details of how such capabilities are implemented in neural tissues is challenging, not only from the neurobiological view, but also from the neuro- and bio-engineering perspective, which aims at modelling the neural circuits underlying such amazing capabilities, building the corresponding circuits and control systems, and subsequently implementing the resulting architectures in neuro-inspired robots.

Brain dynamics are a paradigm of *complexity*, characterised by interactions between many nonlinear dynamical systems, some in cooperation and others in competition, to give rise to a large number of emerging behaviours. Of course, obtaining reliable models requires detailed knowledge of the corresponding neural circuits, which is not available for vertebrate brains. Under this perspective the task will be easier, the smaller the selected brain for modelling will be. Following this guideline, the authors focused their attention on insects. These creatures are definitely not just simple reflex-based automata. Not all their behaviours are inborn; rather it is becoming more and more evident that most of the capabilities that make insects able to survive in extremely diverse environments are being refined by learning. Exploitation of stored individual experience, from short-term to long-term memories, from simple habituation to complex associations is at the heart of adaptive behaviour. Some key examples are reported in this book.

The recent progress in insect neurobiology, a good portion owed to *Drosophila* neurogenetics and the new precise tools, brings about circuitry with single-neuron precision and at the same time its physiology. This created the basis on which the authors could start a mutual collaboration between basic and applied sciences, aimed at designing key experiments to demonstrate specific behavioural capabilities in insects. Modern tools allow to identify where and how in the insect brain such behaviours are initiated, organised, executed and controlled. Subsequently, by identifying underlying specific nonlinear dynamical systems representing neural processing and learning, the research aimed at designing and implementing nonlinear circuits and systems is able to reproduce those identified behaviours and finally to show the results obtained via simulated or actually working robot prototypes. In regard to the circuits level, some of the authors were involved in the last decade in the design of neuromorphic integrated electronic circuits for the real-time generation and control of insect-inspired walking machines. Recently, however, this approach was temporarily abandoned, to leave time for simulation experiments and high-level hardware implementation of the models derived by biological experiments. This was due to the need for flexibility, since often, new results from neurobiology continuously require changes in the architecture realised so far. The approach to modelling and realisation of neural circuits and systems for bio-inspired robot control follows the connectionistic approach, supported by novel tools from neurogenetics. The latter provide the unique chance to address specific functionalities, at least in some cases, down to the single-neuron level in unpreceded short time. Nevertheless, neurobiology and simulation/implementation live on two different timescales, because it takes time to find out how a nervous system actually solved a particular sub-aim, whereas the problem becomes only obvious by and during the modelling process. Reasonable assumptions have to be made, and sometimes the model has to be refined later according to new results from biology. The neurobiologist likes to thank his fellow authors for not having given up to stay as close as possible to the biological role model. That way mutual benefit for basic and applied sciences could be created. The approach, briefly recalled in Chap. 1, is a unique opportunity to model and implement the same capabilities in robots as they are shown by the biological counterpart.

The book concisely presents this emergent research topic which aims at exploring the essence of neural processing in insect brains and consequently at modelling, implementing and realising the corresponding emergent features in artificial bio-inspired robots. It is the result of a multidisciplinary research collaboration established in the course of the last six years. The book starts with a brief review on the main areas of the *Drosophila* brain, describing the current knowledge about their structure and function. Then some of the main behaviours are described that involve learning and memory and are of potential interest to robotics. Chapter 2 describes the main elements which will be used in all of the following chapters of the book to build computational control models: basic neural (spiking and nonspiking) models, synaptic models and learning algorithms are briefly reviewed. Nonlinear dynamical structures, made up of the introduced models, are also dealt with at the end of the chapter and used to show key features like locomotion, motor learning, memory formation and exploitation. Chapter 3 is devoted to computational models of spatial memory formation, in relation to a particular part of the insect brain presented in Chap. 1. The model was simulated, and the corresponding results are reported. Additional complex behaviours such as sequence learning and perception, recently discovered to exist in insects, are also addressed. Important issues, like motor learning and body-size learning, are described in Chap. 4, whereas sequence and sub-sequence learning are reported in Chap. 5, using a model which is the result of a further refinement of an architecture able to reproduce, concurrently, a number of different behaviours: decision making (classification), sequence learning, motor learning, delayed-match to sample, and others, reported elsewhere, like attention and expectation. Chapter 6, taking into account the model introduced in Chap. 5, reviews a typical case of multimodal sensory integration in insects, where several sensory modalities of the insect brain use the same neural path to elicit common motor actions, that way avoiding duplication, but rather sharing neural circuits. This example is discussed in parallel to the concept of *neural reuse*. It is shown how the model, discussed in Chap. 5, is a computational example of such concept in action: the same dynamical core can serve different behavioural needs concurrently. This research field joins together different expertise and is believed to substantially contribute to the advancement of science and engineering.

Catania, Italy   Luca Patanè
Mainz, Germany   Roland Strauss
Catania, Italy   Paolo Arena
September 2017

Acknowledgements

The results reported in the book would have been never reached without the contribution of the European Commission, that, under the FP6 and FP7 projects such as SPARK, SPARK II and EMICAB, substantially contributed to create a multidisciplinary and international research knot in the area of bio-inspired robotics.

Contents

1 Biological Investigation of Neural Circuits in the Insect Brain   1
   1.1 Introduction   1
   1.2 Memories in Orientation   1
   1.3 The *Drosophila* Genetic Construction Kit   4
   1.4 Mushroom Bodies—Structure and Function   5
   1.5 Central Complex—Structure and Function   10
   1.6 Towards Long-Term Autonomy—More Things to be Learned
       from Insects   15
   1.7 Conclusions   16
   References   16

**2 Non-linear Neuro-inspired Circuits and Systems: Processing
and Learning Issues**   21
   2.1 Introduction   21
   2.2 Spiking Neural Models   22
       2.2.1 Leaky Integrate-and-Fire Model   22
       2.2.2 Izhikevich's Neural Model   23
       2.2.3 Resonant Neurons   23
   2.3 Synaptic Models   24
       2.3.1 Synaptic Adaptation Through Learning   24
       2.3.2 Synaptic Model with Facilitation   25
   2.4 The Liquid State Network   25
       2.4.1 Learning in the LSN   26
   2.5 Non-spiking Neurons for Locomotion   27
   2.6 Conclusions   28
   References   29

3 Modelling Spatial Memory   31
   3.1 Introduction   31
   3.2 Ellipsoid Body Model   32

3.3   Model Simulations   36
3.4   Conclusions   42
References   42

4   Controlling and Learning Motor Functions   45
4.1   Introduction   45
4.2   Working Hypotheses for the Model   47
4.3   Simulation Results   49
4.4   Motor Learning in a Climbing Scenario   50
4.5   Body-Size Model   57
4.6   Conclusions   62
References   63

5   Learning Spatio-Temporal Behavioural Sequences   65
5.1   Introduction   65
5.2   Model Structure   68
       5.2.1   Antennal Lobe Model   68
       5.2.2   Initial Model of the MB Lobes   69
       5.2.3   Premotor Area   71
       5.2.4   Context Layer   72
5.3   MB-Inspired Architecture: A Step Ahead   74
       5.3.1   Network Behaviour for Classification   75
       5.3.2   End Sequence Neurons   77
       5.3.3   Neural Models and Learning Mechanisms   77
       5.3.4   Decision-Making Process   78
       5.3.5   Learning Sequences and Sub-sequences   79
5.4   Robotic Experiments   80
5.5   Conclusions   83
References   83

6   Towards Neural Reusable Neuro-inspired Systems   87
6.1   Introduction   87
6.2   Multimodal Sensory Integration   89
6.3   Orientation: An Example of Neural Reuse in the Fly Brain   92
       6.3.1   Role of CX in Mechanosensory-Mediated
               Orientation   92
       6.3.2   Olfactory and Mechanosensory-Based Orientation   93
6.4   Concluding Remarks   96
References   98

Chapter 1
Biological Investigation of Neural Circuits in the Insect Brain

1.1 Introduction

On closer inspection one cannot but adore the behavioural capabilities of insects. First and foremost one might think of social insects like bees and ants with their ability to communicate and organize states, however, the solitary species have developed amazing reproductive, foraging or orientation strategies as well. While mankind evolved one of the largest brains in the animal kingdom, insects followed the path of miniaturization and splendid sparseness. Those features turn them into role models for autonomous robots. Among insects the fruit fly *Drosophila melanogaster* is the most intensely studied species owing to the unrivalled genetic tools available. Despite their tiny brains, fruit flies can orient with respect to outside stimuli—and remember previous encounters with such stimuli, can walk on uneven terrain—and this in any orientation to gravity, can fly in adverse winds, can find partners, places for egg laying, food and shelter. In order to achieve these goals, *Drosophila melanogaster* can build various memories starting from a 4-s visual working memory [36, 52] and a place memory [53], to a minutes- to 3-hours-long idiothetic working memory [55, 62, 63], a short-term and long-term olfactory memory (over days; review: [58]), a courtship-suppression memory [71], all the way to a life-long body-size memory [34]. Flies can also learn at what time during the day certain information is important [10, 11].

1.2 Memories in Orientation

Place learning (cold spot on a hot plate). The paradigm adapts Morris water maze in which swimming rodents remember the position of an invisible platform underneath the water surface of a circular tank with milky water with the help of visual landmarks in the room [50]. The insect adaptation of the paradigm uses a hot plate on which

© The Author(s) 2018
L. Patanè et al., *Nonlinear Circuits and Systems for Neuro-inspired Robot Control*,
SpringerBriefs in Nonlinear Circuits, https://doi.org/10.1007/978-3-319-73347-0_1

a cold spot with pleasant temperature can be found and remembered by visual cues on the walls of the arena. The insect paradigm was first described by Mizunami et al. [48, 49] and used to study orientation in cockroaches. The recent adaptation for *Drosophila* by Ofstad et al. [53] uses a tiled floor of Peltier elements and a circular LED arena to show a 360° panorama of 120° vertical, 120° horizontal and 120° oblique stripes. Several flies can be video-tracked at the same time. The cold spot and the pattern can be switched congruently to alternative positions almost instantaneously after the flies have found the cold spot. A heat barrier and clipped wings keep the flies within the arena and the time until the flies find the cold spot and their track lengths are measured. Flies use visual information to memorize the pleasant place.

Place learning (heat box). The heat-box paradigm was developed by Wustmann et al. [94] to device a large-scale mutant screening for operant conditioning. Flies learn to avoid one half of a long, narrow chamber, because it heats up on entry. Flies learn from their own actions and later stay in the pleasant half even when the heat is turned off. Two different memory components were identified in immediate retention tests, a spatial preference for one side of the chamber and a "stay-where-you-are-effect" [62]. Intermittent training is shown to give higher retention scores than continuous training and strengthens the latter effect. When flies were trained in one chamber and tested in a second one after a brief reminder training, an aftereffect of the training can still be observed 2 h later. The longest memory of 3 h is achieved at aversive temperatures of 41 °C [55]. The various memory effects are independent of the mushroom bodies. Since training and test occur in the dark, and chambers can be switched, the paradigm tests for idiothetic orientation.

Visual orientation memory. In this paradigm single flies with clipped wings walk on a circular platform towards a landmark on a cylindrical LED screen beyond a water-filled moat [52]. This landmark disappears while at the same instance an indistinguishable distracter landmark appears under 90° to the first-seen. One second after the fly has changed its heading towards the distracter, it disappears as well so that the fly is left without visual orientation cues. Normal flies nevertheless return to their original heading and continue the approach of the first-chosen, now invisible landmark. The paradigm tests for a 4-s visual working memory important to bridge phases of interrupted visibility in a cluttered environment. It has been used to identify the seat of the memory in the brain and the underlying biochemistry [36].

Olfactory memory. In the olfactory conditioning paradigm of Tully and Quinn [83] up to 50 flies in a test-tube like structure are confronted with an odour while at the same time electric shocks are given to their feet. Thereafter, a second odour is presented without foot-shocks. Then, the flies are gently pushed into an elevator and brought to a choice point, from where they can walk into two test-tube like chambers facing each other. One chamber presents the previously punished odour, the other the neutral odour. This so-called one-cycle training leads to the formation of a labile memory that can be detected for up to 7 h. Learning is tested within 2 min after training, short-term memory between 30 min and 1 h after training, and middle-term memory 3 h after training. But flies can also acquire olfactory long-term memory that lasts up to one week if they are repeatedly trained with pauses between the training cycles (spaced training). Gene regulation and protein synthesis are inevitable for long-term memory. If the same overall training time as in spaced training is given without pauses, this is called massed training. Massed training leads to the formation of anaesthesia-resistant memory and is assessed one day after training [29, 84]. Appetitive olfactory memory is achieved when sugar reward instead of electric shocks are applied [58, 67]. To allow for a direct comparison between olfactory and visual conditioning, Vogt et al. [89] have developed a four-quadrant arena with transparent electric grid that can be used to administer electric shocks in visual and olfactory experiments.

Courtship suppression. Mated female flies reject courting male flies as long as sperm remains in their spermatheca. Males are actively kicked away and the ovipositor is protruded [71]. While courting, naïve male flies learn to avoid mated females in an aversive conditioning procedure. The taste receptor GR32a functions as pheromone receptor and conveys courtship suppression towards mated females [47]. Delay time to courting virgin females can serve as indictor for learning.

Body-size memory. Certain behavioural decisions depend on body reach, in vertebrates represented in the brain as peripersonal space. E.g. in flies, the decision to overcome a chasm in the walkway by gap-climbing depends on body size, which, in turn, is determined by genetic and environmental factors like temperature and quality of food during larval stages. Therefore, body size is learned after hatching using a multisensory integration strategy. Visual input is correlated with tactile experience or corollary discharge for walking. During acquisition of the own body size, the fly generates parallax motion by walking in a structured environment. The average step size, which is proportional to leg lengths and therefore to body size, creates an average parallax motion on the eyes that allows to calibrate behavioural decisions [34]. Later decisions to climb a gap are instructed by the parallax motion generated on the eyes by edges of the opposite side [60, 82].

Time-of-day memory. In a study by Chouhan et al. [10] starved flies were trained in the morning to associate a particular odour with sucrose reward. A respective training was repeated in the afternoon but with a different odour. The procedure was repeated the next day, and time-dependent odour preference was tested in the morning and afternoon of the third day. *Drosophila* chose the previously rewarded odour dependent on the time of the test whenever the two different training events on day 1 and 2 had been separated by more than 4 h. Flies can form time-odour associations in constant darkness as well as in daily light-dark cycles, but not under constant light-on conditions. The study demonstrates that flies can utilize temporal information as an additional cue in appetitive learning. Indeed, key genes of the circadian clock were essential for time-odour learning in flies. Circadian clock mutants, *period^{01}* and *clock^{AR}*, learned to associate sucrose reward with a certain odour but were unable to form time-odour associations. Those require a *period*- and *clock*-dependent endogenous mechanism. In a follow-up study by Chouhan et al. [11] the extent of starvation was shown to be correlated with the fly's ability to establish time-odour memories. Prolonged starvation promotes time-odour associations after just one single cycle of training. Starvation is required for acquisition but not for retrieval of time-odour memory.

1.3   The *Drosophila* Genetic Construction Kit

Drosophila is *the* model animal for geneticists and cutting-edge tools are available to manipulate the nervous system of the fly and study the underpinnings of its behavioural capabilities. Over the years the fruit fly has been literally turned into a genetic construction kit.

First, so called *driver lines* express a xenogeneic transcription factor like GAL4 just in parts of the nervous system [4, 59]. GAL4 has been borrowed from baker's yeast and is *per se* functionless in *Drosophila*. Thousands of driver lines are available from stock centres, each with a specific expression pattern in the nervous system. E.g., expression can be restricted to all olfactory receptor neurons, to all serotonergic neurons (i.e. neurons using serotonin as their neurotransmitter), to all neurons within a particular brain centre or all neurons using a specific molecule; the lines are catalogued and the list of expression patterns is almost endless.

Second, xenogeneic transcription factors like GAL4 are functionless without their specific binding site, which does not exist in *Drosophila*. The binding site for GAL4 in baker's yeast is upstream activating sequence or UAS. Upon binding of GAL4 to UAS the gene behind UAS is transcribed and expressed. Instead of the genuine yeast gene any gene from any organism can be placed behind UAS and the construct be integrated in the *Drosophila* genome. This constitutes an *effector line* which is ideally functionless without GAL4. Again, thousands of effector lines can be found in stock centres. The term construction kit signifies the fact that parental flies from a driver line and an effector line can be crossed with each other so that the offspring possesses driver and effector elements in one fly. In those individuals, and only in neurons within the expression pattern of the GAL4 line, GAL4 binds to UAS and the effector is expressed. Effectors can be reporter genes like green fluorescent protein, or GFP, taken from a bioluminescent jellyfish, so that the GAL4 expression pattern can be visualized [9]. They can be genes encoding molecules for inactivating the neurons in the expression pattern, like tetanus toxin light chain taken from the tetanus bacterium [77]. In this case the chemical synapses cease to function. Effectors can be genes for additional proteins, but also genes for knocking down the RNA building plan for particular molecules that normally would be expressed in the neuron. RNA interference, as the latter technique is called (review: [14]), is the method of choice for probing into the function of biochemical cascades.

Third, effective methods for controlling the *time of expression of an effector* have been developed. E.g., GAL80^{ts} is a repressor of GAL4-controlled expression in yeast. At 18 °C and lower temperatures it binds to GAL4 (that itself is bound to UAS) and represses transcription of the effector gene behind UAS. At 29 °C and higher temperatures GAL80^{ts} detaches from GAL4 and transcription will start [46]. Time-controlled expression of effectors allows for discerning between developmental and acute functional defects caused by the effector. Other methods of expression control have been developed, e.g. a hormone-based system. Expression starts when a particular hormone is administered by special fly food.

In recent years effectors have been developed that by themselves can be controlled by temperature or light. *Shibire^{ts}* encodes for a dominant negative dynamin needed for recycling of vesicles at the rim of chemical presynapses. Upon shifting the ambient temperature to 32 °C the neurons in the expression pattern are silenced within seconds, because existing vesicles are used up and no new vesicles can be formed [33]. Within reasonable limits, the block is reversible by lowering the temperature. Functional analysis requires also artificial activation of neurons and potent methods have been developed to this end. TrpA1 is a temperature-sensitive cation channel naturally occurring in human and animal cells. Expressed under GAL4 control it allows activating neurons by shifting the temperature [19]. Channels are closed at 21 °C and firing gradually increases with temperature up to a maximum at 29 °C. A third example for controllable effectors are light-activated cation channels (e.g. ReaChR; [28]).

1.4   Mushroom Bodies—Structure and Function

Structure. Mushroom bodies are a paired structure of the *Drosophila* brain; they are found in similar form in all insects. Mushroom bodies develop from four neuroblasts per hemisphere, each of which can produce three different intrinsic neuron types called Kenyon cells. The Kenyon cells developing from *Drosophila* embryo up to and including the second larval stage are γ neurons. Within the third larval stage exclusively α′-/β′-mushroom body neurons are formed. All Kenyon cells born in the subsequent pupal stage are α-/β-neurons [37]. The three Kenyon-cell types refer to the specific axonal projection patterns in the adult stage. All Kenyon cells, about 2000 per hemisphere, receive their main and predominantly olfactory input in the calyx of their brain side. From there they project through the peduncle into one of five mushroom-body lobes. α-/β-Kenyon cells and α'-/β'-Kenyon cells bifurcate at the anterior ventral end of the peduncle and project into the α-/β-lobes (1000 cells; [6]) or α'-/β'-lobes (370 cells), respectively. About 670 γ-Kenyon cells project without bifurcation just into the γ-lobes [6, 12]. α- and α'-lobes are also called vertical lobes, whereas β, β' and γ lobes are referred to as medial lobes (Fig. 1.1).

Laminar organization of the Kenyon cell axons divides the peduncle into at least five concentric strata (Fig. 1.1; [78]). The α-, β-, α'- and β'-lobes are each divided into three strata, whereas the γ lobe appears homogeneous. Specifically the outermost stratum of the α-/β-lobes is connected with the accessory calyx, a protruded subregion of the calyx, which does not receive direct olfactory input [78].

Noteworthy, three cell types have been identified, which are mushroom-body intrinsic but non-Keyon-cell neurons. One pair of MB-DPM neurons (DPM refers to the position of the cell bodies in the dorsal posterior medial area of the brain), one pair of MB-APL neurons (APL: anterior paired lateral), and about 50 MB-AIM neurons (AIM: anterior inferior medial; [78]). The arborisation of the single DPM neuron per hemisphere covers all lobes and the anterior part of the peduncle. The arborisation of the single APL neuron per hemisphere is even wider and comprises,

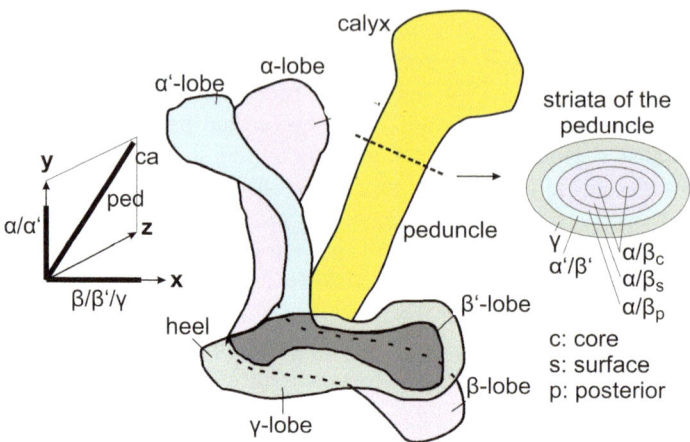

Fig. 1.1 Left mushroom body of *Drosophila melanogaster*. Frontal view. About 2000 Kenyon cells project from the calyx through the peduncle to the five lobes. They either bifurcate into α- and β-lobes, or α'- and β'-lobes, or project without bifurcation into the γ-lobe. The most anterior γ-lobe is shown in grey. The β'-lobe is slightly dorsal to the β-lobe. The α'-lobe winds around the α-lobe, their ends are in the same plane. Inset right: schematic cross section of the peduncle at the dashed line showing the strata of the peduncle (after [78])

in addition to all lobes, also the calyx and the entire peduncle. In contrast, the 25 AIM neurons per hemisphere arborize in small areas of the γ- and β'-lobes [78].

All Kenyon cells of one brain hemisphere converge onto 34 mushroom body output neurons (MBONs) of which 21 types have been discriminated [1]. Judged by their roles in several associative learning tasks and in sleep regulation, some distinct single MBON output channels and a multi-layered MBON network have been identified.

Besides the olfactory input to the calyx region by projection neurons, 17 other types of mushroom body extrinsic neurons (MBENs) were identified by Tanaka et al. [78]. Those arborize in the calyx, lobes, and peduncle. Lobe-associated MBENs arborize in specific areas of the lobes along their longitudinal axes. This fact defines three longitudinal segments in the α- and three in the α'-lobes, two segments in the β- and two in the β'-lobes, and five segments in the γ-lobes. Taken together, the laminar arrangement of the Kenyon cell axons in the lobes and the segmented organization of the MBENs divide the lobes into smaller synaptic units, possibly facilitating characteristic interaction between intrinsic and extrinsic neurons in each unit for different functional activities (schematic maps are given e.g. in [27, 78]).

Olfactory learning. The foremost function of mushroom bodies is found in olfactory learning and memory. Flies naturally learn and can be trained to be attracted by or to avoid particular odours. Odour perception and processing is essential for *Drosophila*, because flies detect food and analyse sex partners using this sense. Olfactory receptors reside on the antennae and the maxillary bulbs; they are localized in the sensory dendrites of olfactory receptor neurons (ORN) that are housed in hair-like, chemosensory sensilla. Each ORN expresses only one type of olfactory receptor, together with the co-receptor Or83b; 62 different primary olfactory receptors are known in *Drosophila*. ORN axons project to the paired antennal lobes, each receptor type determines a particular target glomerulus. In most respects antennal lobes resemble the olfactory bulbs of vertebrates, which are the first stage of olfactory processing, as well [90, 91]. Within their glomerulus, ORNs become connected to projection neurons (PNs) and local interneurons. About 150 cholinergic PNs convey the processed information to the mushroom bodies and the lateral horn [44, 78]. By mushroom-body inactivation experiments it has been determined that the direct influence of the lateral horn is sufficient to particularly release inborn behaviours [22], whereas the mushroom body path serves plasticity of behaviour. More recent studies characterized an additional olfactory path that uses inhibitory PNs projecting exclusively to the lateral horn thereby conveying hedonic value and intensity of odours [76].

Mushroom bodies build associative memories for natural odours. The associated experience may be "good" or "bad" and is used to control adaptive behaviour. A high selectivity for bouquets of odours is achieved, because every Kenyon cell receives input from a randomly developed set of, on average, three PNs [6] and thus serves as a coincidence detector. A high odour selectivity and at the same time low population activity of the mushroom body Kenyon cells is achieved by GABAergic attenuation. "Sparse coding" is fostered by a pair of GABAergic anterior paired lateral neurons (APLs), which innervate the entire mushroom body. The Kenyon cells, in turn, possess $GABA_A$ receptors which allow for chloride anion influx when opened [38]. Altogether, the system allows discrimination between and storage of experience with natural odours for a wide range of primary odorant combinations.

Input from other sensory modalities is anatomically not obvious in *Drosophila* but mushroom-body dependent tasks related to vision are nevertheless described for flies (e.g. [40, 64, 79]). The mushroom bodies of honeybees receive, additional to olfactory information, prominent visual [18], gustatory, and mechanosensory [66] inputs. These connections likely provide mixed-modality signals that lead to experience-dependent structural changes documented in freely behaving bees (e.g. [15]). In flies and bees the lobe regions receive information on sugar reward and electric foot shock. Whereas dopaminergic neurons undisputedly convey negative reinforcement information to Kenyon cells, octopaminergic neurons have long been considered to convey reward to Kenyon cells in insects (e.g. [20, 67]). Burke et al. [5] showed in *Drosophila* that only short-term appetitive memory is reinforced by octopaminergic neurons and that these neurons signal on a specific subset of dopaminergic neurons, which in turn synapse on Kenyon cells. More specifically, hedonic information (sweet taste) goes to the β'_2- and γ_4-regions of the mushroom body lobes and can establish just a short-term memory. Appetitive long-term memory is nutrient-dependent and requires different dopaminergic neurons that project to the γ_{5b}-lobe region, and it can be artificially reinforced by dopaminergic neurons projecting to the β-lobe and adjacent α_1-lobe region [27].

Thus, during associative learning dopaminergic neurons convey the reinforcing effect of the unconditioned stimuli of either valence to the axons of *Drosophila* mushroom body Kenyon cells for normal olfactory learning. Cervantes-Santoval et al. [7] show that Kenyon cells and dopaminergic neurons form axo-axonic reciprocal synapses. The dopaminergic neurons receive cholinergic input from the Kenyon cells via nicotinic acetylcholine receptors. The neurogenetic knock-down of these receptors in dopaminergic neurons impaired olfactory learning. When the synaptic output of the Kenyon cells was blocked during olfactory conditioning, this reduced the presynaptic calcium transients in the dopaminergic neurons, a finding consistent with reciprocal communication and positive feedback. The blocking also reduced the normal chronic activity of the dopaminergic neurons.

The different mushroom body lobes serve distinct functions in the establishment of olfactory memory. The γ-lobes play the central role in short-term memory, whereas the α'-/β'-lobes are indistinguishable for memory consolidation after training. The longer the time lag between training and memory retrieval gets, the more important will be the α-/β-lobes; they are important for the long-term memory [8, 80]. Shortly later came the finding that, within a given Kenyon cell type, particular sub-groups of neurons might be responsible for the engrams of different durability (shell- and core units; [57]).

Using optogenetic artificial activation of MBONs Aso et al. [1] could induce repulsion or attraction in flies. The behavioural effects of MBON activation was combinatorial, suggesting that the MBON ensemble represents valence collectively. Aso et al. propose that local, stimulus-specific dopaminergic modulation selectively alters the balance within the MBON network for those stimuli in order to bias memory-based action selection.

As pointed out in Sect. 1.2, long-term olfactory memory requires spaced training and relies on new protein synthesis in order to stabilize learning-induced changes of synapses. Associations ought to be reproducible, before a long-term memory should form. Wu et al. [93] describe a sequence of events in the mushroom bodies that lead to stable aversive long-term memory from earlier labile phases. In a neurogenetic survey through all 21 distinct types of MBONs Wu and colleagues show that sequential synthesis of learning-induced proteins occurs within just three types of MBONs. Down-regulation of protein synthesis in any of these three MBON types impaired long-term memory. Moreover, synaptic outputs from all three MBON types are all required for memory retrieval. The requirement for protein synthesis during long-term memory consolidation is sequential; first it is needed in MBON-α3 (0–12 h), then in MBON-γ3,γ3β'1, and finally in MBON-β'2mp. The time shift in the requirement for protein synthesis in the second MBON is 2 h and in the third another 2 h.

But even consolidated long-term memory can be extinct if the learned prediction is later inaccurate. If still correct, it will be maintained by re-consolidation. Felsenberg et al. [16] identified neural operations underlying the re-evaluation of appetitive olfactory memory in *Drosophila*. Both, extinction or re-consolidation activate specific parts of the mushroom body output network and specific subsets of reinforcing dopaminergic neurons. Extinction requires MBONs with dendrites in the α- and α'-lobes of the mushroom body, which drive negatively reinforcing dopaminergic neurons that innervate neighbouring compartments of the α- and α'-lobes. The aversive valence of this extinction memory neutralizes the previously learned odour preference. Memory re-consolidation, in turn, requires $\gamma_2\alpha_1'$-MBONs. This pathway recruits negatively reinforcing dopaminergic neurons innervating the same γ-lobe compartment and re-engages positively reinforcing dopaminergic neurons to re-consolidate the original reward memory. Thus, a recurrent and hierarchical connectivity between MBONs and dopaminergic neurons enables memory re-evaluation depending on reward-prediction accuracy.

Other functions. But mushroom bodies serve also other, more general and non-olfactory functions. It has been noticed that flies with ablated or inactivated mushroom bodies enhance their spontaneous walking activity as measured over several hours [42]. Within the first 15 min, however, the walking activity is reduced in comparison to intact flies, because their initial arousal phase is missing without mushroom bodies [70]. Mushroom bodies regulate activity with regard to internal stimuli like hunger and outside stimuli indicating, for instance, food [64]. Repeated inescapable stress reduces behavioural activity with all signs of a depression, whereas repeated reward can reactivate flies and make them resilient to stress ([64]; see chapt. 1.6). When courting, mushroom-body-less flies show deficits in detecting female pheromones [65]; they can neither form short-term nor long-term courtship-suppression memory [45]. If unforeseeable obstacles occur, mushroom-body-less flies have problems switching to a new type of behaviour. For instance, at a water-filled moat they keep trying to reach the landmark beyond the water barrier, sometimes for minutes, despite the fact that their wings are clipped [51]. In flight-simulator experiments, decisions upon contradictory visual stimuli mirror the average between the valences of the stimuli, whereas intact flies take clear-cut decisions for one or the other stimulus [79]. Again in tethered flight, it was found that context generalization requires mushroom bodies [40]. Mushroom-body-less flies learned the heat-punished object and the colour of the background whereas intact flies abstracted from the background and learned the dangerous object as such. Mushroom bodies are the centre of sleep control [96]. Motor learning requires the mushroom bodies [31]. This was shown in a study of gap climbing in flies. Reiterative training can lead to a long-term motor memory. According to electrophysiological experiments, Kenyon cells respond to olfactory, visual, tactile and gustatory stimuli. It has been proposed that mushroom bodies are a multimodal sensory integration centre [12]. This has been confirmed in behavioural experiments (e.g. [64, 89]). With their four-quadrant apparatus Vogt et al. [89] showed that both visual and olfactory stimuli are modulated in the same subset of dopaminergic neurons for positive associative memories. Another subset of dopaminergic neurons was found to drive aversive memories of both the visual and olfactory stimuli. With regard to mushroom-body intrinsic neurons, distinct but partially overlapping subsets are involved in visual and olfactory memories. The work of Vogt et al. shows that associative memories are processed by a centralized circuit that receives both visual and olfactory inputs, thus reducing the number of memory circuits needed for such memories.

1.5 Central Complex—Structure and Function

Structure. The central complex resides at the midline of the protocerebral hemi-
spheres of insects. In *Drosophila* it is composed of four neuropilar regions, the largest
of which is the fan-shaped body. Within an anterior depression resides the ellipsoid
body, a perfectly toroid-shaped neuropil. Ventral to the ellipsoid body are the paired
noduli and posterior to the fan-shaped body a neuropil with the shape of a bicycle's
handle bar is found called protocerebral bridge [21]. The central complex neuropils
are composed of columnar elements, which interconnect the four neuropils and are
often output elements of the central complex, and of tangential cells, which mostly
provide input to all or several columnar elements within one neuropil. A third type
of neurons stays intrinsically and connects homologous elements within neuropils.
The conspicuous columnar organization is seen as 18 glomeruli (9 per hemisphere)
in the protocerebral bridge, 16 sectors within the ellipsoid body, and 8 fans of the
fan-shaped body. E.g. Wolff et al. [92] provide a comprehensive catalogue of 17
cell types arborizing in the protocerebral bridge and insights into the anatomical
structure of the four components of the central complex and its accessory neuropils.
Revised wiring diagrams take into account that the bridge comprises 18 instead of
16 glomeruli as previously published. Most recent information on the wiring of the
fan-shaped body can be found in [3], and on the ellipsoid-body wiring in [54].

The conspicuous tangential elements of the ellipsoid body are called ring neurons;
each ring neuron innervates all 16 sectors of the ellipsoid body. Five ring systems
are known to date (R1–R5; R4 is further subdivided into R4m for medial and R4d
for dorsal) [43, 54]. The average number of ring neurons per set is 20–25 per hemisphere and the total number amounts to 200–250 cells [97, supplement]. According
to [43] 37 ± 4 ring neurons of the adult brain are not GABAergic. Ring neurons
are postsynaptic at the bulb (formerly called lateral triangle; [30]) and predominantly presynaptic within the ring. Novel coincident synapses have been found in
the ellipsoid-body ring; these are precise arrays of two independent active zones over
the same postsynaptic dendritic domain [43]. Such arrays are compatible with a role
as coincidence detectors and represent about 8% of all ellipsoid-body synapses in
Drosophila.

The pathways for visual input from the optic lobes, through the anterior optic
tubercles to the bulbs have been recently described in *Drosophila* [54]. Starting from
the development of the ellipsoid body, Omoto et al. describe that all ring neurons are
formed by a single neuroblast lineage. Two other lineages give rise to the neurons
that connect the anterior optic tubercles with the dendrites of the ring neurons in the
bulb and to neurons that connect the medullae of the optic lobes with the anterior
optic tubercles. This anterior visual pathway conveys information on the polarization
pattern of the sky (sky-compass) in locusts and other insects [26]. The two lineages
form two parallel pathways; DALcl1 neurons connect the peripheral ring neurons
and DALcl2 neurons the central ring neurons with the anterior optic tubercles. All
DALcl1/2 neurons respond predominantly to bright objects. Whereas DALcl1 neurons show small and retinotopically ordered receptive fields on the ipsilateral eye,

the DALcl2 neurons possess one large excitatory receptive field on the contralateral eye that they share, and one large inhibitory field on the ipsilateral side. When a bright object enters the ipsilateral field, they become inhibited and they respond with extra excitation when such an object leaves the ipsilateral receptive field. A second visual pathway is currently known just from larger insects. The protocerebral bridge is connected to the posterior optic tubercles, which in turn receive input from the accessory medullae [26].

Columnar sets of neurons interconnect the central-complex neuropils and accessory areas [21, 39, 92]. They usually come in sets in a multiple of eight.

Functions. Fan-shaped body. Learning experiments at the flight simulator revealed visual memory functions of the fan-shaped body in *Drosophila* [41]. Flies recognize previously heat-punished visual objects by parameters such as size, colour or contour orientation, and store respective parameter values for later avoidance. The features are stored independently of the retinal position of the objects on the eyes during learning for later visual pattern recognition. Short-term memory traces of the pattern parameters 'elevation in the panorama' and 'contour orientation' were localized in two different layers of the fan-shaped body. At the time of the study the fan-shaped body was thought to be six-layered, meanwhile nine layers are distinguished [92]. Layers are perpendicular to the vertical columns in horizontal planes of the fan-shaped body.

Protocerebral bridge. The protocerebral bridge is involved in step-size control for speed increase and directional control of walking and gap-climbing [61, 81]. Structural mutants of the protocerebral bridge are walking slowly because their step size remains at a basic level [75]. In contrast, wild-type flies increase their step size with stepping frequency, thereby reaching almost double the maximum speed of bridge-defective flies [73, 74]. Flies turn by keeping step size on the inside of the curve at the basic level, whereas strides on the outside are increased. Stepping frequency is invariant between inside and outside; a model of how the bridge differentially influences step size is provided in [75]. Flies with clipped wings overcome chasms in the walkway by gap climbing, provided the gap is of surmountable width [60]. The distance to the opposite side is determined by the amount of parallax motion distal edges create on the retina during approach. The fly's own body size, which can vary by 15% within a given genetic background, is taken into account (see Sect. 1.2). Energetically costly climbing is thereby restricted to surmountable gaps. It has been observed that bridge-defective mutant flies are losing direction when initiating climbing. At the end of the proximal side of the catwalk their climbing attempts point in all directions, whereas intact flies stay in a small angular corridor that targets the distal side of the gap [81]. In locusts, the protocerebral bridge is shown to hold a map-like representation of the polarized light information of the sky compass [25].

Ellipsoid body. The ellipsoid body plays essential roles in visual pattern memory [56], orientation memory [35, 36, 52] and place learning [53], and thus it is considered to be a centre of visual learning and memory. Pan and colleagues [56] demonstrated in a follow-up study to Liu et al. [41] at the flight simulator that not just neurons in the fan-shaped body, but also a small set of neurons in the ellipsoid body are involved in visual pattern memory. Localized expression of a learning gene in the respective learning mutant revealed, that the rescue of the memory defect could be achieved in either of the central complex neuropils. Knock-down experiments in either structure demonstrated that both were required for visual pattern memory. A test of different visual pattern parameters, such as size of the retinal image, contour orientation, and vertical compactness revealed differential roles of the fan-shaped body and the ellipsoid body for visual pattern memory.

A short-lived visual orientation memory helps flies to bridge phases of an object approach, during which the chosen target gets temporarily out of sight. Even after a detour to a distracter landmark, flies can return to their previously planned path, given that the distracting landmark is seen for just a short time [52]. The working memory resides in the R3-ring neurons of the ellipsoid body [35]. The 16 sectors are considered to represent azimuth sections in the outside world. Two gaseous neurotransmitters are found, which can explain the volatility of the memory that lasts for about 4 s. One of them, nitric oxide, was known to act as retrograde neurotransmitter from post- to pre-synapses; in orientation memory it acts within R3-ring neurons in a signalling cascade that leads to the production of the second messenger cGMP and likely to the opening of cyclic nucleotide gated cation channels [36]. A second gaseous messenger, hydrogen sulfide, was known to act synergistically with nitric oxide in smooth muscle relaxation, but it was surprising to find the same interaction of the two gases in neurons. The memory trace is modelled as elevated cGMP levels in single sectors of R3 neurons (Fig. 1.2); levels are jointly regulated by the production of nitric oxide, which activates cGMP-producing guanylyl cyclase, and hydrogen sulfide, which inhibits phosphodiesterase 6 that would otherwise degrade cGMP. In accord with the outcome of behavioural experiments, the model assumes that NO and cGMP are formed in the ellipsoid body sector that represents the azimuth sector of the visual field holding the target landmark. When this landmark disappears and a distracter appears at a different azimuth position, the respective sector in the ellipsoid body starts to generate nitric oxide and cGMP, whereas in the original sector the signal starts to degrade. If the distractor is shown for longer than 4 s, it will become the actual target, because its cGMP signal is now stronger than that of the original target landmark (Fig. 1.2). The memory is read out only, if no visual cues are present; it is a back-up system that helps flies to stay on track during short phases of missing visual input. Idiothetic input (e.g. corollary discharge from turning commands) helps to update the orientation memory during such phases by path integration. R2- and/or R4-ring neurons are suitable to provide map-like visual input [68]. The R3-ring neurons provide an off-signal, when the target disappears from sight [54]. Several types of neurons are known [92] that connect the ellipsoid body with the protocerebral bridge for a read-out that differentially influences step size and thus walking direction.

The place memory for a cool spot on a hot surface [53] uses visual cues in the surrounding to re-identify the pleasant place. R1-neurons of the ellipsoid body are required for proper function of the place memory. Of note, the visual information in this paradigm is always available, whereas in the case of the visual orientation mem-

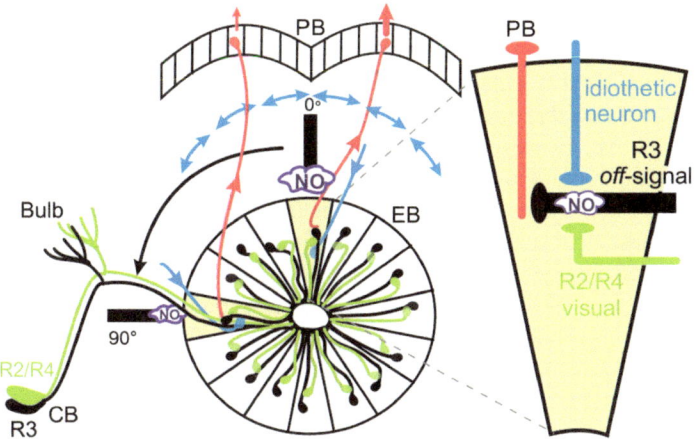

Fig. 1.2 Proposed biological model of visual working memory storage and retrieval. Depicted is the connectivity and flow of information in *Drosophila* in one of the 16 segments of the ellipsoid body (EB) and between the EB and the protocerebral bridge (PB). Ring neurons R3 (black) receive visual input from the R2- or R4-system (green) and simultaneously from neurons conveying information on the fly's self-motion (idiothetic; blue). Landmark approach stimulates NO production in one axonal branch of R3 neurons, and elevated levels of NO induce cGMP production. cGMP-mediated opening of cyclic-nucleotide-gated ion channels (CNGs; violet) results in calcium influx, representing the memory trace of the landmark. When the distractor appears instead of the original target, the described processes are repeated in an EB segment representing the respective azimuth angle. When no object is visible (off signal), a columnar neuron (red) projecting from the EB to the PB will be activated by the R3 neurons. Differential activation of the EB-PB neurons in the segments is then used to steer the fly. Note that a longer-shown distractor will become the main target as its NO/cGMP signal rises, whereas that of the original landmark volatilizes (adapted from [36])

ory R3-neuron activity is required only after no visual cues are left. Thus different ring-neuron systems have distinct orientation functions.

In what form does visual information enter the central complex? To this end Seelig and Jayaraman [68] performed two-photon calcium imaging experiments on behaving flies walking stationary on an airstream-supported sphere. They describe R2- and R4d-ring neurons to be visually responsive. Their dendrites in the accessory area called bulb [30] are retinotopically arranged. Their receptive fields comprise excitatory as well as inhibitory subfields and, according to Seelig and Jayaraman [68], resemble properties of simple cells in the mammalian primary visual cortex V1. Authors found a strong orientation tuning in the R2- and R4d-ring neurons which in some cases was direction-selective. Vertically oriented visual objects evoked particularly strong responses.

In their follow-up study Seelig and Jyaraman [69] demonstrate that visual landmark orientation and angular path integration are combined in the population responses of columnar ellipsoid body neurons whose dendrites tile the ellipsoid body. Authors point out that the responses of these cells are reminiscent of mammalian head direction cells that represent the direction of heading relative to a landmark and update their directional tuning by path integration if the lights are turned off. Calcium peaks were seen to rotate within the ellipsoid body together with the landmark on a virtual-reality panorama. The angular relation between the landmark in the visual world and the position of the calcium peak varied from trial-to-trial, it is therefore relative. However, within a given measuring episode it faithfully represents the angular azimuth motion of the landmark. Moreover, in the absence of visual cues, its rotation represents self-motion of the fly. Several features of the population dynamics of these neurons and their circular anatomical arrangement are suggestive of ring attractors, network structures that have been proposed to support the function of navigational brain circuits. This idea has been elaborated by Kim et al. [32]. Noteworthy, the calcium-imaging experiments showed, that for several landmarks in the environment only one bump, apparently for the chosen landmark, was visible [69].

Turner-Evans et al. [85] and Green et al. [17] present the working of the internal compass orientation within the ellipsoid body of *Drosophila*. In addition to the head-direction cells [85], which project from the ellipsoid body to the protocerebral bridge and an accessory region called gall [30], shifting cells are described. Head-direction cells are systematically termed EBw.s-PBg.b-gall.b neurons by [92], have previously been referred to as eb-pb-vbo neurons by [21], E-PG neurons by [17], EIP neurons by [39], and "wedge neurons" in a review by [23]. They may be homologous to CL1a neurons in the locust [25] and the butterfly [24]. Shifting neurons or P-EN neurons [17] are systematically termed PBG2–9.s-EBt.b-NO1.b neurons [92], correspond to PEN neurons of [39] and tile neurons of [23]. These columnar neurons of the central complex project from the protocerebral bridge to the ellipsoid body and the noduli. While head-direction cells represent the heading in relation to a landmark (and two of them signal from one sector of the ellipsoid body to two individual glomeruli on either side of the protocerebral bridge), the shifting cells become active in conjunction with body rotations of the fly. Two sets of shifting cells are projecting from the identical glomerulus of the bridge to a sector adjacent to where the currently active head-direction cells emerge. The set of shifting cells emerging from the left-hemisphere glomeruli of the bridge is offset by one sector of the ellipsoid body to the right and is responsible for right turns of the body and the calcium signal. The set of shifting cells emerging on the right side of the bridge is offset by one sector to the left and is responsible for left-shifts and counter-clockwise body turns [23].

Last not least it should be noted that the ellipsoid body is involved in certain memory phases of olfactory aversive memory. This came as a surprise, as the mushroom bodies seemed sufficient for a long time to explain olfactory memory. It seems to be a modulation function of the ellipsoid body more than an actual memory trace that is required for aversive olfactory middle-term memory in *Drosophila* [98]. The artificial activation of R2 and R4m neurons did not affect learning with regard to middle-term memory, but eliminated a particular labile component, the anaesthesia-sensitive memory. The majority of the activated ring neurons is inhibitory, thus it seems that they actively suppress the anaesthesia-sensitive memory. Evidence is provided, that ellipsoid body ring neurons and mushroom body medial lobes may be synaptically connected.

1.6 Towards Long-Term Autonomy—More Things to be Learned from Insects

Autonomy in roving robots is currently mostly restricted to autonomous orientation and locomotion in rather short-term goals defined by humans. Admittedly, autonomous vacuum cleaners and lawn mowers show a first sign of a "long-term survival strategy", when they reach for their power station upon their batteries going low. But on Mars, robots will need more of such long-term strategies. Just two thoughts on that.

Robots might have to find new resources and that calls for curiosity behaviour, whenever the robot is in a relaxed and saturated state. Insects show novelty choice behaviour, which can be seen as one component of curiosity. It requires memory of what is already familiar in order to identify new objects in the environment. Those objects are consequently explored. In flight-simulator experiments involving tethered-flying fruit flies, Dill and Heisenberg [13] were the first to show that flies preferentially oriented for objects with a new size or a new visual contrast with regard to the background. van Swinderen and colleagues argue that novelty comprises an aspect of visual salience in *Drosophila*. In local field potentials derived from the fly's central brain, novelty of a stimulus increases endogenous 20-to-30-Hz local field potential activity [86–88]. According to Solanki et al. [72], three functional components underlie novelty choice at the flight simulator, visual azimuth orientation, a working memory, and the ability for pattern discrimination. In an attempt to map novelty choice in the fly brain, they find that both the central complex and the mushroom bodies are involved. The latter are specifically needed when it comes to a comparison of patterns of different sizes. Within the central complex particularly the ellipsoid body and its ring-neuron systems are involved.

A compact system for activity control is another feature we might learn from insects. Ries et al. [64] recently found an activity control system in *Drosophila*. It resides in the mushroom bodies and uses serotonin as neurotransmitter and neurohormone. Hungry flies become more active and even more, if food odour is provided. Long-term stress, on the other hand, in the study consisting of inescapable sequences of adverse vibrations that were unforeseeably repeated over days, causes a depletion of serotonin (5-HT) in the mushroom bodies; the depletion correlates with behavioural inactivity with all signs of depression. Sugar reward, in turn, replenishes 5-HT in the mushroom bodies and leads to normal activity. The bipartite system requires tightly balanced serotonergic signalling to two different lobes of the mushroom bodies to generate adequate behavioural activity. Activation of α-lobes (relevant Kenyon cells express 5-HT-1A receptors) enhances behavioural activity, whereas activation of γ-lobes (relevant Kenyon cells express 5-HT-1B receptors) reduces such activity. Modern antidepressants suppress reuptake of 5-HT from the synaptic cleft so that 5-HT molecules activate postsynaptic receptors more intensely; they ameliorate the depression-like state of *Drosophila*, too. The relevant next questions for basic research are now, how a simple nervous system adds up stress information and how it identifies reward. Short episodes of stress activate rather than depress the fly. It is the long-term repetition of stress, and the fact that it is inescapable, that leads to activity reduction (for learned helplessness in flies see [2, 95]). A similar integration happens with reward: the normal amount of sugars in the fly food has no antidepressant effect. A short sugar rush, however, given after each day of vibration stress and before the normal food overnight, turns flies resilient to stress. While the signalling path of externally-administered reward is understood to some extent, we need to find out how a simple nervous system can deduce reward-like signals from successful own actions as opposed to punishment-like signals from unsuccessful own attempts.

1.7 Conclusions

The tremendous progress in neurogenetic tools has advanced our understanding of *Drosophila* brain functions. Mushroom bodies and central complex are in the centre of interest of basic research and precise information on the actual wiring of behavioural control circuits is coming up. Calcium imaging is the method of choice when it comes to watching the brain at work, but the dynamics the method can mirror is restricted. Electrophysiological studies are needed to complement imaging. Modelling can guide basic researchers to their next experiments, but it can also be used to develop bio-inspired control of autonomously roving robots. This book wants to show what we can learn from *Drosophila* and other insects in order to achieve such astounding adaptive behaviour that flies control with just about 150000 central neurons.

References

1. Aso, Y., Sitaraman, D., Ichinose, T., Kaun, K.R., Vogt, K., Belliart-Guérin, G., Plaçais, P.Y., Robie, A.A., Yamagata, N., Schnaitmann, C., Rowell, W.J., Johnston, R.M., Ngo, T.T., Chen, N., Korff, W., Nitabach, M.N., Heberlein, U., Preat, T., Branson, K.M., Tanimoto, H., Rubin, G.M.: Mushroom body output neurons encode valence and guide memory-based action selection in *Drosophila*. eLife **3**, e04580 (2014)
2. Batsching, S., Wolf, R., Heisenberg, M.: Inescapable stress changes walking behavior in flies-learned helplessness revisited. PLoS ONE **11**(11), e0167066 (2016)
3. Boyan, G., Liu, Y., Khalsa, S.K., Hartenstein, V.: A conserved plan for wiring up the fan-shaped body in the grasshopper and *Drosophila*. Dev. Genes Evolut. **227**(4), 253–269 (2017)
4. Brand, A.H., Perrimon, N.: Targeted gene expression as a means of altering cell fates and generating dominant phenotypes. Development **118**(2), 401–415 (1993)
5. Burke, C.J., Huetteroth, W., Owald, D., Perisse, E., Krashes, M.J., Das, G., Gohl, D., Silies, M., Certel, S., Waddell, S.: Layered reward signaling through octopamine and dopamine in *Drosophila*. Nature **492**(7429), 433–437 (2012)
6. Caron, S.J.C., Ruta, V., Abbott, L.F., Axel, R.: Random convergence of olfactory inputs in the *Drosophila* mushroom body. Nature **497**(7447), 113–117 (2013)

7. Cervantes-Sandoval, I., Phan, A., Chakraborty, M., Davis, R.L.: Reciprocal synapses between mushroom body and dopamine neurons form a positive feedback loop required for learning. eLife **10**, e23789 (2017)
8. Cervantes-Sandoval, I., Martin-Pena, A., Bery, J., Davis, R.: System-like consolidation of olfactory memories in *Drosophila*. J. Neurosci. **33**(23), 9846–9854 (2013)
9. Chalfie, M., Tu, Y., Euskirchen, G., Ward, W.W., Prasher, D.C.: Green fluorescent protein as a marker for gene expression. Science **263**(5148), 802–805 (1994)
10. Chouhan, N.S., Wolf, R., Helfrich-Förster, C., Heisenberg, M.: Flies remember the time of day. Curr. Biol. **25**(12), 1619–1624 (2015)
11. Chouhan, N.S., Wolf, R., Heisenberg, M.: Starvation promotes odor/feeding-time associations in flies. Learn. Mem. **24**(7), 318–321 (2017)
12. Crittenden, J., Skoulakis, E., Han, K., Kalderon, D., Davis, R.: Tripartite mushroom body architecture revealed by antigenic markers. Learn. Mem. **5**(1–2), 38–51 (1998)
13. Dill, M., Heisenberg, M.: Visual pattern memory without shape recognition. Philos. Trans. R. Soc. Lond. B **349**, 143–152 (1995)
14. Duffy, J.B.: GAL4 system in *Drosophila*: a fly geneticist's Swiss army knife. Genesis **34**(1–2), 1–15 (2002)
15. Farris, S.M., Robinson, G.E., Fahrbach, S.E.: Experience-and age-related outgrowth of intrinsic neurons in the mushroom bodies of the adult worker honeybee. J. Neurosci. **21**(16), 6395–6404 (2001)
16. Felsenberg, J., Barnstedt, O., Cognigni, P., Lin, S., Waddell, S.: Re-evaluation of learned information in *Drosophila*. Nature **544**, 240–244 (2017)
17. Green, J., Adachi, A., Shah, K.K., Hirokawa, J.D., Magani, P.S., Maimon, G.: A neural circuit architecture for angular integration in *Drosophila*. Nature **546**, 101–106 (2017)
18. Gronenberg, W., López-Riquelme, G.O.: Multisensory convergence in the mushroom bodies of ants and bees. Acta Biol. Hung. **55**(1), 31–37 (2004)
19. Hamada, F.N., Rosenzweig, M., Kang, K., Pulver, S., Ghezzi, A., Jegla, T.J., Garrity, P.A.: An internal thermal sensor controlling temperature preference in *Drosophila*. Nature **454**, 217–220 (2008)
20. Hammer, M., Menzel, R.: Multiple sites of associative odour learning as revealed by local brain microinjections of octopamine in honeybees. Learn. Mem. **5**, 146–156 (1998)
21. Hanesch, U., Fischbach, K.-F., Heisenberg, M.: Neuronal architecture of the central complex in *Drosophila melanogaster*. Cell Tissue Res. **257**, 343–366 (1989)
22. Heimbeck, G., Bugnon, V., Gendre, N., Keller, A., Stocker, R.: A central neural circuit for experience-independent olfactory and courtship behavior in *Drosophila melanogaster*. Proc. Natl. Acad. Sci. USA **98**(26), 15336–15341 (2001)
23. Heinze, S.: Neural coding: bumps on the move. Curr. Biol. **27**, R410–R412 (2017)
24. Heinze, S., Reppert, S.M.: Sun compass integration of skylight cues in migratory monarch butterflies. Neuron **69**, 345–358 (2011)
25. Heinze, S., Gotthardt, S., Homberg, U.: Transformation of polarized light information in the central complex of the locust. J. Neurosci. **29**, 11783–11793 (2009)
26. Homberg, U.: Sky compass orientation in desert locusts-evidence from field and laboratory studies. Front. Behav. Neurosci. **9**, 346 (2015)
27. Huetteroth, W., Perisse, E., Lin, S., Klappenbach, M., Burke, C., Waddell, S.: Sweet taste and nutrient value subdivide rewarding dopaminergic neurons in *Drosophila*. Curr. Biol. **25**(6), 751–758 (2015)
28. Inagaki, H.K., Jung, Y., Hoopfer, E.D., Wong, A.M., Mishra, N., Lin, J.Y., Tsien, R.Y., Anderson, D.J.: Optogenetic control of freely behaving adult *Drosophila* using a red-shifted channelrhodopsin. Nat. Methods **11**(3), 325–332 (2014)
29. Isabel, G., Pascual, A., Preat, T.: Exclusive consolidated memory phases in *Drosophila*. Science **304**, 1024–1027 (2004)
30. Ito, K., Shinomiya, K., Ito, M., Armstrong, D., Boyan, G., Hartenstein, V., Harzsch, S., Heisenberg, M., Homberg, U., Jenett, A., Keshishian, H., Restifo, L.L., Rössler, W., Simpson, J.H., Strausfeld, N.J., Strauss, R., Vosshall, L.B.: A systematic nomenclature of the insect brain. Neuron **81**, 755–765 (2014)

31. Kienitz, B.: Motorisches Lernen in Drosophila Melanogaster. Shaker Verlag, Aachen (2010)
32. Kim, S.S., Rouault, H., Druckmann, S., Jayaraman, V.: Ring attractor dynamics in the *Drosophila* central brain. Science **356**(6340), 849–853 (2017)
33. Kitamoto, T.: Conditional modification of behavior in *Drosophila* by targeted expression of a temperature-sensitive *shibire* allele in defined neurons. J. Neurobiol. **47**, 81–92 (2001)
34. Krause, T., Strauss, R.: Body reach learning from parallax motion in *Drosophila* requires PKA/CREB. J. Neurogenet. **26** Suppl. 1, 48 (2012)
35. Kuntz, S., Poeck, B., Sokolowski, M.B., Strauss, R.: The visual orientation memory of *Drosophila* requires Foraging (PKG) upstream of Ignorant (RSK2) in ring neurons of the central complex. Learn. Mem. **19**, 337–340 (2012)
36. Kuntz, S., Poeck, B., Strauss, R.: Visual working memory requires permissive and instructive NO/cGMP signaling at presynapses in the *Drosophila* central brain. Curr. Biol. **27**(5), 613–623 (2017)
37. Lee, T., Lee, A., Lou, L.: Development of the *Drosophila* mushroom bodies: sequential generation of three distinct types of neurons form a neuroblast. Development **126**(18), 4065–4076 (1999)
38. Lei, Z., Chen, K., Li, H., Liu, H., Guo, A.: The GABA system regulates the sparse coding of odors in the mushroom bodies of *Drosophila*. Biochem. Biophys. Res. Commun. **436**(1), 35–40 (2013)
39. Lin, C.-Y., Chuang, C.-C., Hua, T.-E., Chen, C.-C., Dickson, B.J., Greenspan, R.J., Chiang, A.-S.: A comprehensive wiring diagram of the protocerebral bridge for visual information processing in the *Drosophila* brain. Cell Rep. **3**, 1739–1753 (2013)
40. Liu, L., Wolf, R., Ernst, R., Heisenberg, M.: Context generalization in *Drosophila* visual learning requires the mushroom bodies. Nature **400**(6746), 753–756 (1999)
41. Liu, G., Seiler, H., Wen, A., Zars, T., Ito, K., Wolf, R., Heisenberg, M., Liu, L.: Distinct memory traces for two visual features in the *Drosophila* brain. Nature **439**(7076), 551–556 (2006)
42. Martin, J., Ernst, R., Heisenberg, M.: Mushroom bodies suppress locomotor activity in *Drosophila melanogaster*. Learn. Mem. **5**(1), 179–191 (1998)
43. Martin-Pena, A., Acebes, A., Rodriguez, J.-R., Chevalier, V., Triphan, T., Strauss, R., Ferrus, A.: Cell types and coincident synapses in the ellipsoid body of *Drosophila*. Eur. J. Neurosci. **39**(10), 1586–1601 (2014)
44. Masse, N., Turner, G., Jefferis, G.: Olfactory information processing in *Drosophila*. Curr. Biol. **19**(16), R700–R713 (2009)
45. McBride, S., Giuliani, G., Choi, C., Krause, P., Correale, D., Watson, K., Baker, G., Siwicki, K.: Mushroom body ablation impairs short-term memory and long-term memory of courtship conditioning in *Drosophila melanogaster*. Neuron **24**(4), 967–977 (1999)
46. McGuire, S.E., Le, P.T., Osborn, A.J., Matsumoto, K., Davis, R.L.: Spatiotemporal rescue of memory dysfunction in *Drosophila*. Science **302**(5651), 1765–1768 (2003)
47. Miyamoto, T., Amrein, H.: Suppression of male courtship by a *Drosophila* pheromone receptor. Nat. Neurosci. **11**(8), 874–876 (2008)
48. Mizunami, M., Weibrecht, J., Strausfeld, N.: A new role for the insect mushroom bodies: place memory and motor control. In: Beer, R. (ed.) Biological Neural Networks in Invertebrate Neuroethology and Robotics, pp. 199–225. Academic Press, Cambridge (1993)
49. Mizunami, M., Weibrecht, J., Strausfeld, N.: Mushroom bodies of the cockroach: their participation in place memory. J. Comp. Neurol. **402**, 520–537 (1998)
50. Morris, R.: Spatial localization does not require the presence of local cues. Learn. Motiv. **12**, 239–260 (1981)
51. Mronz, M., Strauss, R.: Proper retreat from attractive but inaccessible landmarks requires the mushroom bodies. In: 9th European Symposium on Drosophila Neurobiology, Neurofly Dijon (Abstract) (2002)
52. Neuser, K., Triphan, T., Mronz, M., Poeck, B., Strauss, R.: Analysis of a spatial orientation memory in *Drosophila*. Nature **453**(7199), 1244–1247 (2008)
53. Ofstad, T.A., Zuker, C.S., Reiser, M.B.: Visual place learning in *Drosophila melanogaster*. Nature **474**(7350), 204–207 (2011)

54. Omoto, J.J., Keleş, M.F., Nguyen, B.-C.M., Bolanos, C., Lovick, J.K., Frye, M.A., Hartenstein, V.: Visual input to the *Drosophila* central complex by developmentally and functionally distinct neuronal populations. Curr. Biol. **27**(8), 1098–1110 (2017)
55. Ostrowski, D., Kahsai, L., Kramer, E.F., Knutson, P., Zars, T.: Place memory retention in *Drosophila*. Neurobiol. Learn. Mem. **123**, 217–224 (2015)
56. Pan, Y., Zhou, Y., Guo, C., Gong, H., Gong, Z., Liu, L.: Differential roles of the fan-shaped body and the ellipsoid body in *Drosophila* visual pattern memory. Learn. Mem. **16**, 289–295 (2009)
57. Perisse, E., Yin, Y., Lin, A., Lin, S., Hütteroth, W., Waddell, S.: Different Kenyon cell populations drive learned approach and avoidance in *Drosophila*. Neuron **79**(5), 945–956 (2013)
58. Perisse, E., Burke, C., Huetteroth, W., Waddell, S.: Shocking revelations and saccharin sweetness in the study of *Drosophila* olfactory memory. Curr. Biol. **23**(17), R752–R763 (2013)
59. Pfeiffer, B.D., Jenett, A., Hammonds, A.S., Ngo, T.-T.B., Misra, S., Murphy, C., Scully, A., Carlson, J.W., Wan, K.H., Laverty, T.R., Mungall, C., Svirskas, R., Kadonaga, J.T., Doe, C.Q., Eisen, M.B., Celniker, S.E., Rubin, G.M.: Tools for neuroanatomy and neurogenetics in *Drosophila*. Proc. Natl. Acad. Sci. USA **105**, 9715–9720 (2008)
60. Pick, S., Strauss, R.: Goal-driven behavioral adaptations in gap-climbing *Drosophila*. Curr. Biol. **15**, 1473–1478 (2005)
61. Poeck, B., Triphan, T., Neuser, K., Strauss, R.: Locomotor control by the central complex in *Drosophila*-an analysis of the *tay bridge* mutant. Dev. Neurobiol. **68**, 1046–1058 (2008)
62. Putz, G., Heisenberg, M.: Memories in *Drosophila* heat-box learning. Learn. Mem. **9**(5), 349–359 (2002)
63. Putz, G., Bertolucci, F., Raabe, T., Zars, T., Heisenberg, M.: The S6KII (*rsk*) gene of *Drosophila melanogaster* differentially affects an operant and a classical learning task. J. Neurosci. **24**(44), 9745–9751 (2004)
64. Ries, A.-S., Hermanns, T., Poeck, B., Strauss, R.: Serotonin modulates a depression-like state in *Drosophila* responsive to lithium treatment. Nat. Commun. **8**, 15738 (2017)
65. Sakai, T., Kitamoto, T.: Differential roles of two major brain structures, mushroom bodies and central complex, for *Drosophila* male courtship behavior. J. Neurobiol. **66**(8), 821–834 (2006)
66. Schröter, U., Menzel, R.: A new ascending sensory tract to the calyces of the honeybee mushroom body, the subesophageal-calycal tract. J. Comp. Neurol. **465**(2), 168–178 (2003)
67. Schwärzel, M., Monastirioti, M., Scholz, H., Friggi-Grelin, F., Birman, S., Heisenberg, M.: Dopamine and octopamine differentiate between aversive and appetitive olfactory memories in *Drosophila*. J. Neurosci. **23**, 10495–10502 (2003)
68. Seelig, J.D., Jayaraman, V.: Feature detection and orientation tuning in the *Drosophila* central complex. Nature **503**, 262–266 (2013)
69. Seelig, J.D., Jayaraman, V.: Neural dynamics for landmark orientation and angular path integration. Nature **521**, 186–191 (2015)
70. Serway, C., Kaufman, R., Strauss, R., de Belle, J.: Mushroom bodies enhance initial motor activity in *Drosophila*. J. Neurogenet. **23**(1–2), 173–184 (2009)
71. Siegel, R.W., Hall, J.C.: Conditioned responses in courtship behavior of normal and mutant *Drosophila*. Proc. Natl. Acad. Sci. USA **76**, 3430–3434 (1979)
72. Solanki, N., Wolf, R., Heisenberg, M.: Central complex and mushroom bodies mediate novelty choice behavior in *Drosophila*. J. Neurogen. **29**(1), 30–37 (2015)
73. Strauss, R., Heisenberg, M.: Coordination of legs during straight walking and turning in *Drosophila melanogaster*. J. Comp. Physiol. A **167**, 403–412 (1990)
74. Strauss, R., Hanesch, U., Kinkelin, M., Wolf, R., Heisenberg, M.: *No bridge* of *Drosophila melanogaster*: portrait of a structural mutant of the central complex. J. Neurogen. **8**, 125–155 (1992)
75. Strauss, R., Krause, T., Berg, C., Zäpf, B.: Higher brain centers for intelligent motor control in insects. In: Jeschke, S., Liu, H., Schilberg, D. (eds.) ICIRA 2011, Part II, LNAI 7102, pp. 56–64. Springer, Berlin (2011)
76. Strutz, A., Soelter, J., Baschwitz, A., Farhan, A., Grabe, V., Rybak, J., Knade, M., Schmucker, M., Hansson, B., Sachse, S.: Decoding odor quality and intensity in the *Drosophila* brain. eLife **3**, e04147 (2014)

77. Sweeney, S.T., Broadie, K., Keane, J., Niemann, H., O'Kane, C.J.: Targeted expression of tetanus toxin light chain in *Drosophila* specifically eliminates synaptic transmission and causes behavioral defects. Neuron **14**, 341–351 (1995)
78. Tanaka, N.K., Tanimoto, H., Ito, K.: Neuronal assemblies of the *Drosophila* mushroom body. J. Comput. Neurosci. **508**, 711–755 (2008)
79. Tang, S., Guo, A.: Choice behavior of *Drosophila* facing contradictory visual cues. Science **294**(5546), 1543–1547 (2001)
80. Trannoy, S., Redt-Clouet, C., Dura, J., Preat, T.: Parallel processing of appetitive short- and long-term memories in *Drosophila*. Curr. Biol. **21**(19), 1647–1653 (2011)
81. Triphan, T., Poeck, B., Neuser, K., Strauss, R.: Visual targeting of motor actions in climbing *Drosophila*. Curr. Biol. **20**, 663–668 (2010)
82. Triphan, T., Nern, A., Roberts, S.F., Korff, W., Naiman, D.Q., Strauss, R.: A screen for constituents of motor control and decision making in *Drosophila* reveals visual distance-estimation neurons. Sci. Rep. **6**, 27000 (2016)
83. Tully, T., Quinn, W.G.: Classical conditioning and retention in normal and mutant *Drosophila melanogaster*. J. Comp. Physiol. A **157**(2), 263–277 (1985)
84. Tully, T., Preat, T., Bonyton, S.C., Del Vecchio, M.: Genetic dissection of consolidated memory in *Drosophila*. Cell **79**, 35–47 (1994)
85. Turner-Evans, D., Wegener, S., Rouault, H., Franconville, R., Wolff, T., Seelig, J.D., Druckmann, S., Jayaraman, V.: Angular velocity integration in a fly heading circuit. eLife **6**, e23496 (2017)
86. van Swinderen, B.: Attention-like processes in *Drosophila* require short-term memory genes. Science **315**, 1590–1593 (2007)
87. van Swinderen, B., Greenspan, R.J.: Salience modulates 20–30 Hz brain activity in *Drosophila*. Nat. Neurosci. **6**, 579–586 (2003)
88. van Swinderen, B., McCartney, A., Kauffman, S., Flores, K., Agrawal, K., Wagner, J., Paulk, A.: Shared visual attention and memory systems in the *Drosophila* brain. PLoS ONE **4**, e5989 (2009)
89. Vogt, K., Schnaitmann, C., Dylla, K.V., Knapek, S., Aso, Y., Rubin, G.M., Tanimoto, H.: Shared mushroom body circuits underlie visual and olfactory memories in *Drosophila*. eLife **3**, e02395 (2014)
90. Vosshall, L., Stocker, R.: Molecular architecture of smell and taste in *Drosophila*. Annu. Rev. Neurosci. **30**, 505–533 (2007)
91. Vosshall, L., Wong, A., Axel, R.: An olfactory sensory map in the fly brain. Cell **102**(2), 147–159 (2000)
92. Wolff, T., Iyer, N.A., Rubin, G.M.: Neuroarchitecture and neuroanatomy of the *Drosophila* central complex: a GAL4-based dissection of protocerebral bridge neurons and circuits. J. Comp. Neurol. **523**, 997–1037 (2015)
93. Wu, J.-K., Tai, C.-Y., Feng, K.-L., Chen, S.-L., Chen, C.-C., Chiang, A.-S.: Long-term memory requires sequential protein synthesis in three subsets of mushroom-body output neurons in *Drosophila*. Sci. Rep. **7**, 7112 (2017)
94. Wustmann, G., Rein, K., Wolf, R., Heisenberg, M.: A new paradigm for operant conditioning of *Drosophila melanogaster*. J. Comp. Physiol. A **179**(3), 429–436 (1996)
95. Yang, Z., Bertolucci, F., Wolf, R., Heisenberg, M.: Flies cope with uncontrollable stress by learned helplessness. Curr. Biol. **23**(9), 799–803 (2013)
96. Yi, W., Zhang, Y., Tian, Y., Guo, J., Li, Y., Guo, A.: A subset of cholinergic mushroom body neurons requires go signaling to regulate sleep in *Drosophila*. Sleep **36**(12), 1809–1821 (2013)
97. Young, J.M., Armstrong, J.D.: Structure of the adult central complex in *Drosophila*: Organization of distinct neuronal subsets. J. Compar. Neurol. **518**, 1500–1524 (2010)
98. Zhang, Z., Li, X., Guo, J., Li, Y., Guo, A.: Two clusters of GABAergic ellipsoid body neurons modulate olfactory labile memory in *Drosophila*. J. Neurosci. **33**(12), 5175–5181 (2013)

Chapter 2
Non-linear Neuro-inspired Circuits and Systems: Processing and Learning Issues

2.1 Introduction

To model neuro-inspired circuits and systems, we need basic blocks that can be combined with each other, to develop complex networks. In the following chapters, we will present neuro-inspired models designed to solve tasks that range from the control of locomotion to the learning of spatial memory and behavioural sequences. Depending on the peculiar characteristics of the modelled system, different basic processing units were considered. In the following sections, all the ingredients needed will be introduced and discussed.

As far as the neural models are considered, we can identify two elementary processing units: spiking and non-spiking neurons. Spiking neuron models are used to develop bio-inspired neural networks that try to mimic the information transfer that occurs in the brain which is mainly based on spiking events. The use of spikes improves the robustness of the network to noise because the spike can be more easily propagated through neurons even under low signal-to-noise conditions. Networks based on spiking neurons can perform different types of computations from classification to working-memory formation and sequence discrimination [1, 17, 32]. To connect spiking neurons with each other, different synaptic models can be considered [12, 27]. The synaptic transfer function can be either a simple weight or a more complex system with one or several state variables. Furthermore, the parameters of the synaptic model can be subject to a learning process because it is important to store the acquired knowledge coming from the environment in the neural structure [6, 9].

Even if the spiking activity is the most common way to exchange information between in-vivo neurons, there are also examples of non-spiking neurons that regulate through slow dynamics specific rhythmic activities like in the mollusk *Clione* [10]. They are also common in insect nervous systems, mainly to regulate motor activity [13, 29]. These neurons work like pulse generators and can also maintain a plateau when needed. Even in presence of spiking neurons, there is the possibility to sum the

© The Author(s) 2018
L. Patanè et al., *Nonlinear Circuits and Systems for Neuro-inspired Robot Control*,
SpringerBriefs in Nonlinear Circuits, https://doi.org/10.1007/978-3-319-73347-0_2

activity of a group of spiking/bursting neurons using higher level indicators as for instance the mean firing rate that is a quantitative variable able to indicate the level of activity in a population of neurons. This strategy, exploited in several works, is part of the so-called *neural field* approach [18, 24].

The design of the basic blocks, together with the topology of the neural networks are relevant aspects to be addressed also in view of a software/hardware implementation of the architectures. Several solutions could be taken into account to develop neuromorphic circuits [14, 20, 34]: microcontrollers, FPGA, FPAA, integrated circuits and GPU accelerators are some of the potential devices that can be successfully considered.

2.2 Spiking Neural Models

Among the different models available for modelling spiking neurons, we considered different options: the leaky integrate-and-fire (LIF), the Izhikevich's model and the Morris-Lecar resonant neuron.

2.2.1 Leaky Integrate-and-Fire Model

The leaky integrate-and-fire model is one of the first computational models applied to develop neuro-inspired networks [15, 33]. Its main advantage consists of the simplicity of the equations. In fact, the system works like a leaky integrator. The time evolution of the membrane potential $V_m(t)$ of each neuron is described by the following equation:

$$C_m \dot{V}_m(t) = -g_L(V_m(t) - V_L) - I_{syn}(t) \tag{2.1}$$

where C_m is the membrane capacitance (typical values adopted are: 0.5 nF for excitatory cells and 0.2 nF for inhibitory cells); g_L is the leak conductance (0.025 μS for excitatory cells and 0.02 μS for inhibitory cells), and V_L is the resting potential (-70 mV for both excitatory and inhibitory cells). When the membrane potential of a neuron reaches a threshold (i.e. -50 mV) a spike occurs and the membrane potential returns to a reset potential (i.e. -60 mV). The last term $I_{syn}(t)$ represents the total synaptic input current of the cell.

2.2.2 Izhikevich's Neural Model

Izhikevich's neural model, proposed in [22], is well known in the literature and offers many advantages from the computational point of view. The model is represented by the following differential equations:

$$\dot{v} = 0.04v^2 + 5v + 140 - u + I$$
$$\dot{u} = a(bv - u) \tag{2.2}$$

with the spike-resetting

$$\text{if} \quad v \geq 0.03, \quad \text{then} \begin{cases} v \leftarrow c \\ u \leftarrow u + d \end{cases} \tag{2.3}$$

where v is the membrane potential of the neuron, u is a recovery variable and I is the synaptic current. The values assigned to the parameters a, b, c and d will vary as they are depending on the particular neural structures that will be introduced. Varying the neuron parameters we can obtain several different behaviours [22].

2.2.3 Resonant Neurons

In some applications it can be useful to enhance neuron sensitivity to the input stimulus timing or phase. For instance a stimulus, independently of its intensity cannot, to a certain extent, elicit any response if endowed with a given frequency or phase. Neurons able to show such a pattern of activity are called resonators. The Morris-Lecar neural model shows a robust tunable resonant behaviour [19, 31] through the following dynamics:

$$\begin{cases} \dot{V} = k_f[I + g_l(V_t - V) + g_k w(V_k - V) + \\ \quad + g_{Ca} m_\infty(V)(V_{Ca} - V)] \\ \dot{\omega} = k_f[\lambda(V)(\omega_\infty(V) - \omega)] \end{cases} \tag{2.4}$$

where

$$m_\infty(V) = \frac{1}{2}(1 + \tanh \frac{V - V_1}{V_2})$$
$$\omega_\infty(V) = \frac{1}{2}(1 + \tanh \frac{V - V_3}{V_4})$$
$$\lambda(V) = \frac{1}{3} \cosh \frac{V - V_3}{2V_4}$$

Table 2.1 Parameters of the Morris-Lecar neurons

V_1	V_2	V_3	V_4	V_t	V_k
-1.2	18	2	30	-60	-84
g_l	g_k	V_{Ca}	g_{Ca}	I	k_f
2	8	120	4.4	[60.5 61.5]	[0.687 2.5]

V and ω are the state variables of the system, I is the input and V_i, V_t, V_k, V_{Ca}, g_l, g_k, g_{Ca}, I and k_f are parameters of the model whose typical values are indicated in Table 2.1.

2.3 Synaptic Models

Neurons are connected through synapses; the synaptic model transforms the spiking dynamics of the pre-synaptic neuron into a current that excites the post-synaptic one. The mathematical response of the synapses to a pre-synaptic spike can be ruled by the following equation:

$$\varepsilon(t) = \begin{cases} Wt/\tau \exp\left(1 - t/\tau\right), & \text{if } t > 0 \\ 0, & \text{if } t < 0 \end{cases} \tag{2.5}$$

where t is the time elapsed since the emitted spike, τ is the time constant and W is the efficiency of the synapse. This last parameter can be modulated with experience. This model represents the impulse response of the pre-synaptic neuron; it can be cumulated if multiple spikes are emitted within the active window, and in relation to the chosen time constant.

2.3.1 Synaptic Adaptation Through Learning

The Spike-timing-dependent-plasticity STDP can reproduce Hebbian learning in biological neural networks [8, 30]. The algorithm works on the synaptic weights, modifying them according to the temporal sequence of occurring spikes. The updating rule can be expressed by the following formula:

$$\delta W = \begin{cases} A^+ \exp\left(\delta t/\tau^+\right), & \text{if } \delta t < 0 \\ -A^- \exp\left(\delta t/\tau^-\right), & \text{if } \delta t > 0 \end{cases} \tag{2.6}$$

where δt is the time delay between pre- and post-synaptic spikes. In this way the synapse is reinforced if the pre-synaptic spike happens before the post-synaptic one;

it is weakened in the opposite situation. Parameters τ_+ and τ_- represent the slope of exponential functions, whereas positive constants A_+ and A_- represent the maximum variation of the synaptic weight.

2.3.2 Synaptic Model with Facilitation

Short-term synaptic plasticity is characterized by different phenomena and mechanisms. Short-term facilitation is one of these mechanisms that contribute to the evaluation of the synaptic efficacy. A facilitating dynamics can be modeled with the following dynamical system [21]:

$$\dot{u} = (U - u)/F + k\delta(t - t_n) \tag{2.7}$$

where U is a constant which the membrane potential tends to, F is related to the fading memory and k is a gain factor. $\delta(t - t_n)$ represents the spike emitted at time t_n by the pre-synaptic neuron.

2.4 The Liquid State Network

Besides simple networks using spiking neurons, the main structure able to generate neural dynamics important for the models presented in the following chapters is the Liquid state network (LSN). This structure was introduced taking, from the one side, inspiration from specific parts of the insect brain, like mushroom bodies (MBs; see Chap. 1), and on the other side, to already existing architectures and related algorithms. From the latter side, among the different kinds of neural networks used for solving problems like navigation [31], multi-link system control [16] and classification, a lot of interest was devoted to Reservoir computing, which mainly includes two different approaches: Echo State Network (ESN) and Liquid State Machines (LSM) [23, 25]. In previous studies non-spiking Recurrent Neural Networks were used to model the MBs' memory and learning functions [3]. The core of the proposed architecture, inspired by the biology of MBs, resembles the LSM architecture. It consists of a large collection of spiking neurons, the so-called liquid layer, receiving time-varying inputs from external sources as well as recurrent connections from other nodes of the same layer. The recurrent structure of the network turns the time-dependent input into spatio-temporal patterns in the neurons. These patterns are read-out by linear discriminant units. In the last years LSM became a reference point for replicating brain functionalities. However, there is no guaranteed way to analyse the role of each single neuron activity for the overall network dynamics: the control over the process is very weak. This apparent drawback is a consequence of the richness of the dynamics potentially generated within the liquid layer. The side advantage is that the high dimensional complexity can be concurrently exploited through several projections (the read-out maps) to obtain non-linear mappings useful for performing different tasks at the same time.

In details, the architecture used in the following applications consists of a lattice of Izhikevich's class I neurons. An important characteristics of this structure is the local connectivity that is a relevant added value in view of a hardware implementation of the model. Considering that, as recently found in insects, memory-relevant neural connections are excitatory [11], the network used in the following models is composed by excitatory (75%) and inhibitory (25%) neurons. Moreover, in our model the synaptic weight values are randomly distributed between -0.5 and 0.5, whereas the input weights are fixed to 1. The generation of the synaptic connections within the lattice is based on a probability that is a function of the distance $d_{i,j}$ between the presynaptic (i) and postsynaptic (j) neurons.

$$P_{ij} = k * C_{i,j} \tag{2.8}$$

where $C_{inh,inh} = 0.2$, $C_{inh,exc} = 0.8$, $C_{exc,inh} = 0.4$, $C_{exc,exc} = 0.6$. and

$$
\begin{aligned}
k &= 1 \quad \text{if} \quad d_{i,j} \leq 1 \\
k &= 0.5 \ \text{if} \ 1 < d_{i,j} \leq 2 \\
k &= 0 \quad \text{if} \quad d_{i,j} > 2
\end{aligned}
\tag{2.9}
$$

The parameters $C_{i,j}$ have been chosen according to [25]. The distance is calculated considering the neurons distributed on a regular grid with toroidal boundary conditions. The distance $d_{i,j} = 1$ is considered for both horizontally and vertically adjacent neurons. The time constant of the synaptic model was randomly chosen among the values $\tau = 2.5, 5, 15$ and 25 ms. This variability showed to improve the dynamics that can be generated inside the network within the processing time window. The rich dynamics from the LSN is collected into a readout map, whose weights are randomly initialized around zero and are subject to learning. The neurons of the readout map, called Sum neurons, possess a linear activation function and are massively connected with the LSN.

2.4.1 Learning in the LSN

Inputs are provided to the network as currents that, through a sparse connection, reach the hidden lattice (i.e. the liquid layer). The multiple read-out maps can be learned considering the error between the network output, collected through each Sum neuron for each read-out map, and the target signal.

To enslave the dynamics of the LSN to follow a given target, depending on the model needs, a supervised learning method can be performed. A simple solution consists in a batch algorithm, based on the Moore-Penrose Pseudo-inverse computing method, that determines all the weights needed in the read-out map using all the available data in a single iteration. If either the dataset is not completely available since the beginning, or a more biologically grounded solution is preferred, an incremental approach can be adopted.

In this case, for a given sum neuron s, the weight value at each integration step depends on the lattice activity and on the error between the current output and the desired target value. This can be summarized in the following equation:

$$W_{i,j}^s(t+1) = W_{i,j}^s(t) + \eta * Z_{i,j}(t) * E^s(t) \qquad (2.10)$$

where η is the learning factor, $Z_{i,j}(t)$ is the synaptic output of the neuron (i, j) of the lattice at time t and $E^s(t)$ is the error between the desired target and the neuron s. Similar results could be obtained cumulating the error in each presented sample and updating the weights at the end of the presentation. Moreover, to stabilize the learning process, an error threshold $(E_{th} = 10^{-8})$ is imposed to avoid a negligible weight update for each sample presentation.

2.5 Non-spiking Neurons for Locomotion

Non-spiking neurons are perfect candidates for reproducing rhythmic movements that are commonly observed in motor activities. One of the main paradigms of neural networks tailored to work as locomotion controllers is the Central Pattern Generator (CPG) [4].

Among the large number of different implementations of the CPG paradigm, the following architecture will be used in the next chapters.

The basic cell characterizing the CPG network is described by the following equations:

$$\begin{cases} \dot{x}_{1,i} = -x_{1,i} + (i + \mu + \varepsilon)y_{1,i} - s_1 y_{2,i} + i_1 \\ \dot{x}_{2,i} = -x_{2,i} + s_2 y_{1,i} + (i + \mu - \varepsilon)y_{2,i} + i_2 \end{cases} \qquad (2.11)$$

with $y_i = tanh(x_i)$ and the parameters for each cell: $\mu = 0.23$, $\varepsilon = 0$, $s_1 = s_2 = 1$, $i_1 = i_2 = 0$, the system will generate a stable limit cycle [8]. μ is chosen to approximate the dynamics to a harmonic (non-spiking) oscillation. A suitable modulation of these parameters can modify the dynamics of the cell, in order to show also a rhythmic spiking activity. The CPG network is built by connecting neighbouring cells via links expressing rotational matrices $R(\phi)$, as follows:

$$\dot{x}_i = f(x_i, t) + k\Sigma_{j \neq i}(R(\phi_{i,j})x_j - x_i) \text{ with } i, j = 1, \ldots, n \qquad (2.12)$$

where the summation involves all the neurons j which are nearest neighbours to the neuron i; n is the total number of cells; $f(x_i, t)$ represents the reactive dynamics of the i-th uncoupled neurons as reported in Eq. (2.11) and k is the strength of the connections. The sum of these terms performs diffusion on adjacent cells and induces phase-locking as a function of rotational matrices [28].

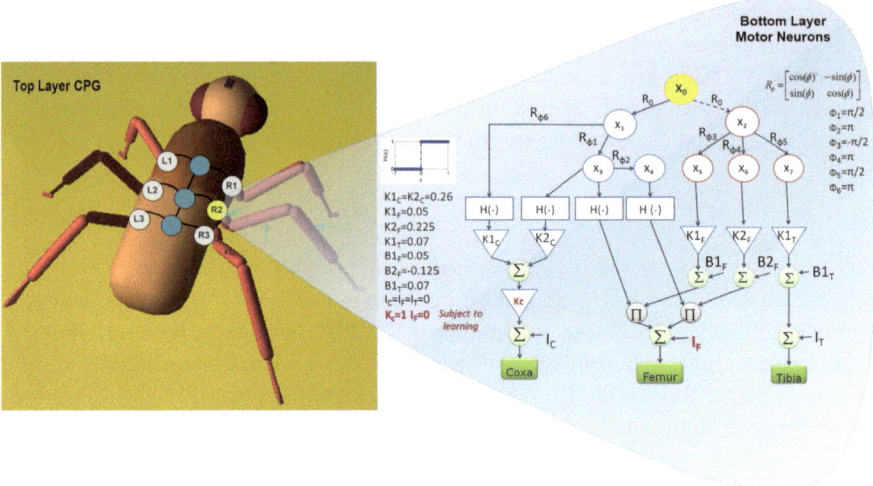

Fig. 2.1 Neural network scheme: the top layer generates a stable gait pattern, whereas the bottom layer consists of additional sub-networks generating the specific reference signals for the leg joints. The network devoted to control a middle leg is reported. The parameters adapted during the learning process for the middle legs are indicated in red [3]

The CPG structure main role is to generate a given stable phase shift among a number of oscillating cells. Of course, an interface needs to be adapted to the peculiarities of the robot kinematics. Moreover, once the robot has walked, each leg has basically two distinct phases: the stance phase (when the leg is on the ground and supports the weight of the body), and the swing phase (when the leg is lifted-off and recovers). In the case of a *Drosophila*-like hexapod simulated robot, the network controlling one of the middle legs is sketched in Fig. 2.1. The CPG neuron identified with the label $R2$ is connected through rotational matrices with different phases to *Top layer CPG*, a network of motor neurons arranged implementing Eq. (2.12) in a directed tree graph, using Eq. (2.11) as the constituting neuron model. The blocks $H(\bullet)$ in Fig. 2.1 are Heaviside functions, used to distinguish, within the limit cycle, between the stance and swing phases: this allows to associate suitable modulation parameters to each part of the cycle, depending on the morphology of the leg. The signals are finally merged to generate the position control command for the coxa, femur and tibia joints. A detailed discussion on the CPG structure and behaviours can be found in a previous study [2].

2.6 Conclusions

In this chapter the ingredients that will be used to develop the neuro-inspired models are introduced. Different neuron models have been considered including spiking and non-spiking ones. Synaptic connections are fundamental to develop networks of neurons. Simple synaptic models have been considered including learning mechanisms that modify the synaptic efficiency depending on the pre- and post- synaptic neuron activity. A first example of complete network was briefly discussed illustrating the learning mechanism needed to enslave the internal dynamics using read-out maps that can be associated to different behavioural responses. Non-spiking neurons finally presented with particular attention to the models used to develop locomotion controllers for bio-inspired robots.

References

1. Abbott, L., DePasquale, B., Memmesheimer, R.: Building functional networks of spiking model neurons. Nat. Neurosci. **19**(3), 350–355 (2016)
2. Arena, E., Arena, P., Patanè, L.: CPG-based Locomotion Generation in a *Drosophila*-inspired Legged Robot. In: Biorob 2012, pp. 1341–1346. Roma, Italy (2012)
3. Arena, E., Arena, P., Strauss, R., Patanè, L.: Motor-skill learning in an insect inspired neuro-computational control system. Front. Neurorobotics **11**, 12 (2017). https://doi.org/10.3389/fnbot.2017.00012
4. Arena, P.: The central pattern generator: a paradigm for artificial locomotion. Soft Comput. **4**(4), 251–265 (2000). www.scopus.com. Cited By :19
5. Arena, P., Caccamo, S., Patanè, L., Strauss, R.: A computational model for motor learning in insects. In: International Joint Conference on Neural Networks (IJCNN), pp. 1349–1356. Dallas, TX (2013)
6. Arena, P., De Fiore, S., Patanè, L., Pollino, M., Ventura, C.: Stdp-based behavior learning on tribot robot. Proceedings of SPIE—The International Society for Optical Engineering, vol. 7365, pp. 1–11 (2009). https://doi.org/10.1117/12.821380
7. Arena, P., Fortuna, L., Frasca, M., Patanè, L.: A CNN-based chip for robot locomotion control. IEEE Trans. Circuits Syst. I **52**(9), 1862–1871 (2005)
8. Arena, P., Fortuna, L., Frasca, M., Patanè, L.: Learning anticipation via spiking networks: application to navigation control. IEEE Trans. Neural Netw. **20**(2), 202–216 (2009)
9. Arena, P., Patanè, L.: Simple sensors provide inputs for cognitive robots. IEEE Instrum. Meas. Mag. **12**(3), 13–20 (2009). https://doi.org/10.1109/MIM.2009.5054548
10. Arshavsky, Y.I., Beloozerova, I.N., Orlovsky, G.N., Panchin, Y.V., Pavlova, G.A.: Control of locomotion in marine mollusc clione limacina iii. on the origin of locomotory rhythm. Exp. Brain Res. **58**(2), 273–284 (1985)
11. Barnstedt, O., David, O., Felsenberg, J., Brain, R., Moszynski, J., Talbot, C., Perrat, P., Waddell, S.: Memory-relevant mushroom body output synapses are cholinergic. Neuron **89**(6), 1237–1247 (2017). https://doi.org/10.1016/j.neuron.2016.02.015
12. Brette, R., Rudolph, M., Carnevale, T., Hines, M., Beeman, D., Bower, J.M., Diesmann, M., Morrison, A., Goodman, P.H., Harris, F.C., Zirpe, M., Natschläger, T., Pecevski, D., Ermentrout, B., Djurfeldt, M., Lansner, A., Rochel, O., Vieville, T., Muller, E., Davison, A.P., El Boustani, S., Destexhe, A.: Simulation of networks of spiking neurons: a review of tools and strategies. J. Comput. Neurosci. **23**(3), 349–398 (2007)
13. Büschges, A., Wolf, H.: Nonspiking local interneurons in insect leg motor control. i. common layout and species-specific response properties of femur-tibia joint control pathways in stick insect and locust. J. Neurophysiol. **73**(5), 1843–1860 (1995). http://jn.physiology.org/content/73/5/1843
14. Chen, Q., Wang, J., Yang, S., Qin, Y., Deng, B., Wei, X.: A real-time FPGA implementation of a biologically inspired central pattern generator network. Neurocomputing **244**, 63–80 (2017). https://doi.org/10.1016/j.neucom.2017.03.028

15. Compte, A., Brunel, N., Goldman-Rakic, P., Wang, X.: Synaptic mechanisms and network dynamics underlying spatial working memory in a cortical network model. Cereb. Cortex **10**, 910–923 (2000)
16. Cruse, H.: MMC—a new numerical approach to the kinematics of complex manipulators. Mech. Mach. Theory **37**, 375–394 (2002)
17. Durstewitz, D., Seamans, J., Sejnowski, T.: Neurocomputational models of working memory. Nat. Neurosci. **19**(3), 1184–1191 (2000)
18. Erlhagen, W., Bicho, E.: The dynamic neural field approach to cognitive robotics. J. Neural Eng. **3**(3), R36 (2006)
19. Hoppensteadt, F., Izhikevich, E., Arbib, M.A. (eds.): Brain Theory and Neural Networks, vol. 181–186, 2nd edn. MIT press, Cambridge (2002)
20. Indiveri, G., Linares-Barranco, B., Hamilton, T., van Schaik, A., Etienne-Cummings, R., Delbruck, T., Liu, S.C., Dudek, P., Hfliger, P., Renaud, S., Schemmel, J., Cauwenberghs, G., Arthur, J., Hynna, K., Folowosele, F., SAGHI, S., Serrano-Gotarredona, T., Wijekoon, J., Wang, Y., Boahen, K.: Neuromorphic silicon neuron circuits. Front. Neurosci. **5**(73), 1–23 (2011). https://doi.org/10.3389/fnins.2011.00073
21. Izhikevich, E., Desai, N., Walcott, E., Hoppensteadt, F.: Bursts as a unit of neural information: selective communication via resonance. TRENDS Neurosci. **26**(3), 161–167 (2003)
22. Izhikevich, E.M.: Which model to use for cortical spiking neurons? IEEE Trans. Neural Netw. **15**(5), 1063–1070 (2004)
23. Jaeger, H.: The "echo state" approach to analysing and training recurrent neural networks. GMD-Report German National Research Institute for Computer Science **148** (2001)
24. Johnson, J.S., Spencer, J.P., Luck, S.J., Schoner, G.: A dynamic neural field model of visual working memory and change detection. Psychol. Sci. **20**(5), 568–577 (2009). https://doi.org/10.1111/j.1467-9280.2009.02329.x
25. Maass, W., Natschlger, T., Markram, H.: Real-time computing without stable states: a new framework for neural computation based on perturbations. Neural Comput. **14**(11), 2531–2560 (2002)
26. Morris, C., Lecar, H.: Voltage oscillations in the barnacle giant muscle fiber. Biophys. J. **35**, 193–213 (1981)
27. Morrison, A., Diesmann, M., Gerstner, W.: Phenomenological models of synaptic plasticity based on spike timing. Biol. Cybern. **98**(6), 459–478 (2008)
28. Seo, K., Slotine, J.: Models for global synchronization in cpg-based locomotion. In: Proceedings 2007 IEEE International Conference on Robotics and Automation, pp. 281–286 (2007)
29. Siegler, M.V.: Nonspiking interneurons and motor control in insects. Adv. Insect Physiol. **18**, 249–304 (1985). https://doi.org/10.1016/S0065-2806(08)60042-9
30. Song, S., Miller, K.D., Abbott, L.F.: Competitive Hebbian learning through spike-timing-dependent plasticity. Nat. Neurosci. **3**, 919–926 (2000)
31. Tani, J.: Model-based learning for mobile robot navigation from the dynamical systems perspective. IEEE Trans. Syst. Man Cybern. Part B **26**(3), 421–436 (1996)
32. Thalmeier, D., Uhlmann, M., Kappen, H.J., Memmesheimer, R.M.: Learning universal computations with spikes. PLOS Comput. Biol. **12**(6), 1–29 (2016). https://doi.org/10.1371/journal.pcbi.1004895
33. Tuckwell, H.: Introduction to Theoretical Neurobiology. Cambridge UP (1988)
34. Wang, R., Cohen, G., Stiefel, K., Hamilton, T., Tapson, J., van Schaik, A.: An fpga implementation of a polychronous spiking neural network with delay adaptation. Front. Neurosci. **7**(14), 1–14 (2013). https://doi.org/10.3389/fnins.2013.00014

Chapter 3
Modelling Spatial Memory

3.1 Introduction

Visual place learning and path integration are relevant capabilities for autonomous robotic systems. Probing into bio-inspired solutions within the animal world, insects like ants and fruit flies can walk in complex environments using different orientation mechanisms: for tracking temporarily obscured targets and for reaching places of interest like a food source, a safe place or the nest. For species that construct nests, the homing mechanisms are fundamental. In the case of desert ants, the route to the nest can be found also after long foraging travels in unstructured environments [18]. In this case, mechanisms of path integration are important to avoid accuracy problems [5]. Landmark navigation can be used to compensate the cumulative errors typical of odometric-based strategies [7].

Neural structures based on mutually coupled populations of excitatory and inhibitory neurons were used to model the navigation behaviour of desert ants [10]. The formation of activity bumps within the neuron populations is used to embed the system orientation in the neural structure. Different mathematical formulations were also considered based on the sinusoidal arrays that condensate the representation of the information using vectors [19]. Besides path integration, interesting approaches for landmark navigation were developed using recurrent neural networks [8] and vision-based strategies that involve population of circular array cells [9].

Together with insects, rats were a useful source of inspiration. In these animals the head direction is codified in cells of the limbic system [15, 17]. While moving in the environment, the animal internally encodes its orientation using a persistent hill of neural activity in a ring-shaped population of excitatory neurons. The position of the activity peak is shifted while turning, using the angular head velocity that is provided as input to other two rings of inhibitory neurons [17].

As previously introduced, insects are able to use spatial information to memorize visual features (spatial distribution, color, etc.) so that they can return to interesting places and can avoid dangerous objects. Furthermore, also insects are able to solve a

© The Author(s) 2018
L. Patanè et al., *Nonlinear Circuits and Systems for Neuro-inspired Robot Control*,
SpringerBriefs in Nonlinear Circuits, https://doi.org/10.1007/978-3-319-73347-0_3

problem similar to the famous Morris water maze problem [14], i.e. an experimental setup where the animal is forced to reach a safe, invisible place in an tank, relying only on external (extra-maze) cues. In insects, the neural circuits responsible for these behaviours need to be further analysed.

The idea to consider *Drosophila melanogaster* as a model organism has been introduced since Chap. 1: it is followed here and in the following chapters. This insect species is particularly interesting for the possibility to apply genetic manipulation tools, to identify the neural processes at the basis of a specific behaviours to be further implemented and demonstrated in bio-inspired robots. Concerning the formation of spatial working memories, even if flies do not create a nest, targeting behaviours are continuously used. Therefore, retaining and recalling a target's position is needed especially when this disappears for a short time.

One experiment used to demonstrate the fly's capabilities in spatial orientation is performed using the detour paradigm where the presentation of a distracter allows to evaluate the robustness of the developed spatial memory also in presence of disturbances [12, 13]. The available genetic manipulation tools identified the important role of the central complex and in particular of the ellipsoid body (EB) in the spatial memory formation process.

In this chapter, on the basis of the neural model proposed in [17] and directly related to orientation in mammals, an adaptation to the insect EB structure has been considered including a further processing level needed for the exploitation of the spatial information contained in the spiking neural structure [2]. Further research led to the discovery of such behaviours as landmark orientation and path integration in specific neural structures within the fruit fly's ellipsoid body [16].

3.2   Ellipsoid Body Model

To model the creation of a spatial working memory in the ellipsoid body, three populations of interconnected neurons have been considered.

We took inspiration from other existing models where a concentration of spiking activity in a part of the network is used to store the heading position of the system acquired through proprioceptive sensors [1, 10].

The model contains one population of excitatory cells ($N_E = 20$) and two populations of inhibitory cells ($N_{I1} = 20$ and $N_{I2} = 20$ neurons). The neuron model considered for the simulation was the Leaky Integrate and Fire whose characteristics were underlined in Chap. 2. The number of neurons considered for the modelling purposes is related to the known neurobiological information on the central complex in *Drosophila* [21] as also briefly discussed in Chap. 1. Neurons in each population are labelled by their heading directions and distributed on a ring that follows the EB circular shape.

The connection weights among neurons depend on their mutual angular positions in the chain. A scheme of the network is reported in Fig. 3.1 where the three neu-

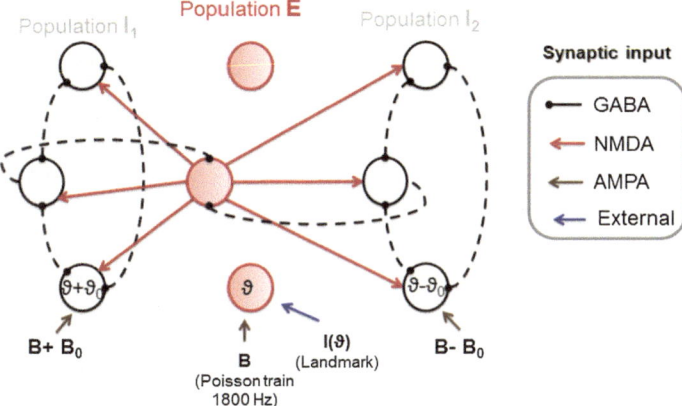

Fig. 3.1 Simple scheme of the EB model: one population of excitatory (E) and two populations of inhibitory (I_1 and I_2) neurons are indicated. Excitatory connections are mediated by AMPA and NMDA receptors whereas GABA is considered for inhibitory connections. In the model a Poisson spike train at 1800 Hz is provided as input to all the network neurons. The presence of a landmark is modelled with a constant input to a given angular position. All-to-all connections between the inhibitory neurons are considered in the model

ral populations together with the connection topology and the external inputs are illustrated.

Each inhibitory neuron has all-to-all connections between other neurons of the same type, with synaptic weights that follow the distribution reported in Fig. 3.2. The excitatory neurons have all-to-all connections with the inhibitory populations with a weight profile reported in Fig. 3.2, whereas each neuron of the inhibitory population is connected with only one neuron of the excitatory population: the neuron that corresponds to the angle $\theta + \theta_0$ in population I_1, inhibits the excitatory neuron that corresponds to the angle θ; this receives also a current contribution from the neuron labeled with $\theta - \theta_0$ in population I_2. In the original model the connection scheme included all-to-all connections also for the interaction between these layers [17]; the simplification presents minimum drawbacks in terms of level of noise in the network as will be presented in the simulations.

Details on the synaptic connections models via NMDA receptors and parameters are reported in [17] in relation to mammals. However, the presence of several receptors, including NMDA [11] was identified in the fly central complex. Moreover the role of NMDA receptors for long-term memory consolidation in the ellipsoid body was assessed [20].

The network receives information on the angular velocity that is integrated through neural processing to provide step-by-step the system orientation.

The angular velocity signal is provided to the network using an uncorrelated Poisson spike train B with basic frequency $f = 1800$ Hz (see Fig. 3.1). Although the spike rate of the input appears to be large, the number represents the total spike input from a group of upstream neurons with a biologically realistic spiking activity.

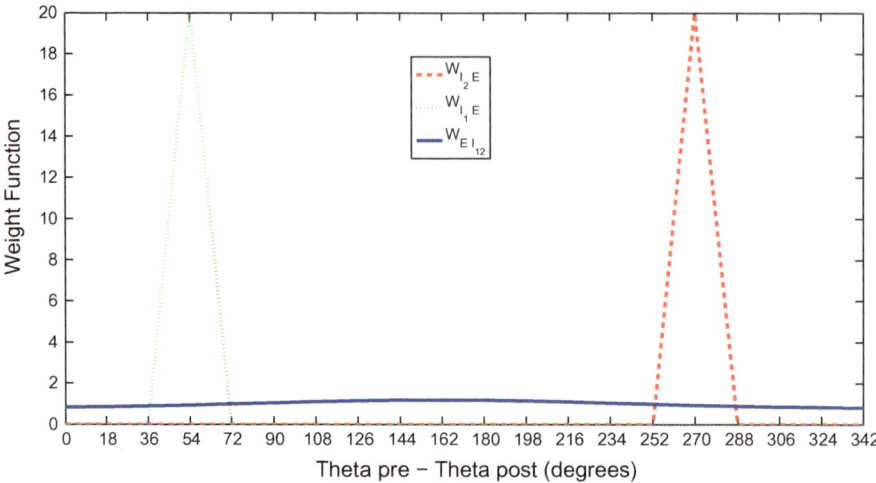

Fig. 3.2 Weight distribution for the synaptic connections involving the neuron population E, I_1 and I_2. This figure was reprinted from [2], © IEEE 2013, with permission

Experiments with other spiking distributions (e.g. regular train) were also carried out obtaining similar results. During a rotation performed by the system, the corresponding angular speed is acquired by the network, giving rise to an unbalance between the input on the two inhibitory populations.

The frequency bias (B_0) used in this case has been characterized through simulations as reported in Sect. 3.3. Finally an external input current (i.e. $I(\theta)$) can be assigned to a neuron in the excitatory population to indicate the presence of a landmark in a specific angular position. The behaviour of the network consists of a shift of the hill of activity to reach the landmark position (Fig. 3.3).

The first experiments performed with the proposed model were designed in absence of sensory and/or self-motion signals: in this case the insect heading is assumed as fixed. In our model, this is implemented by setting to zero ($B_0 = 0\,\text{Hz}$) the head velocity bias of the external afferent inputs provided to the two inhibitory populations. The network, without external stimuli, quickly reaches a steady state condition with a bell-shaped activity profile. The symmetries of the network do not allow to predict where the peak of the activity hill can arise (i.e. depending on the initial conditions, level of noise, etc.). The absence of synaptic connections in the excitatory population indicates that the hill of activity is generated and maintained by the combination of the external excitatory input and the internal inhibitory input coming from the populations I_1 and I_2. When the frequency bias B_0 is positive (negative) a clockwise (anti-clockwise) rotation of the hill of spiking activity is determined.

The presence of visual landmarks is modelled with an input current in the excitatory population in correspondence to the landmark position. The landmark can be also used as a calibration mechanism to shift the hill to a specific position used as

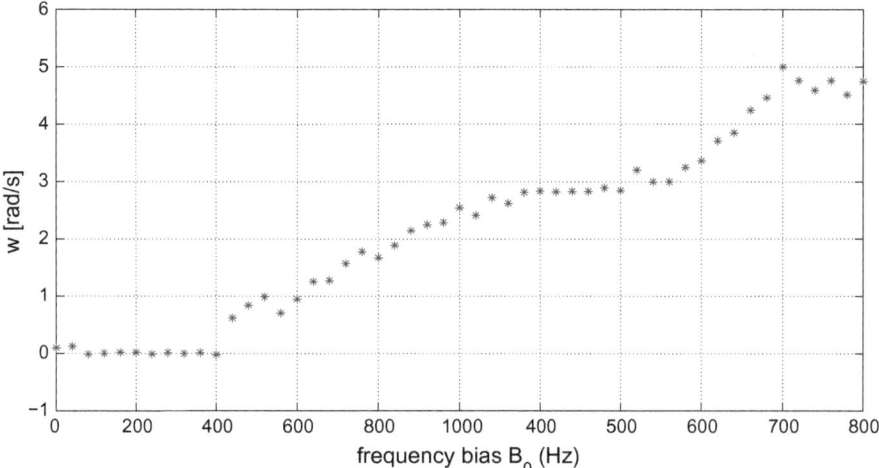

Fig. 3.3 Relation between the frequency bias (B_0) in the Poisson-spike-train and the correspondent angular velocity of the hill of activity in the ring neurons

reference point. During movements the accumulated error can be compensated using visual landmarks placed in known positions.

The developed architecture is able to store the animal's heading direction through a bell-shaped activity profile in the excitatory population. To solve more complex tasks like the detour paradigm [13], a spatial integration of the distance travelled by the agent is required. For the sake of simplicity we consider two basic motion behaviours: forward movement and turning on the spot.

An additional ring of neurons (i.e. population P) has been added to the excitatory population to work as a path integrator. These neurons were modelled with a simple linear transfer function. A scheme of the adapted network is reported in Fig. 3.4 where four distinct populations of neurons are considered. The synaptic connections used to connect population E and population P neurons, are supposed to be characterized by a facilitating dynamics as introduced in Chap. 2:

$$\dot{u}_n = U - u/F + k\delta_n(t - t_s)\delta(B_0) \tag{3.1}$$

where $n = 1..N$, $U = 0$, F is a fading factor and k is a gain related to the system speed, $\delta_n(t - t_s)$ represents the spikes emitted at time t_s by neurons of population E.

A particularity of the proposed model is the presence of a gating function ($\delta(B_0)$) that allows the contribution of a spike at time t_s only if the system is in forward motion (i.e. $B_0 = 0\,\text{Hz}$). When the system is turning, the population P is not activated because the action is performed without spatial translation. The parameter F represents the rate of discharge of the synaptic state variable and has been fixed to a value that allows to retain the information in a time window of about 5 s as also shown in the biological case [13].

Fig. 3.4 Extension of the
network with the
introduction of a fourth ring
needed to perform the detour
experiments. Population P is
used to integrate the forward
motion using a gating
function that mediates the
activity avoiding to consider
turning-on-the-spot
manoeuvres

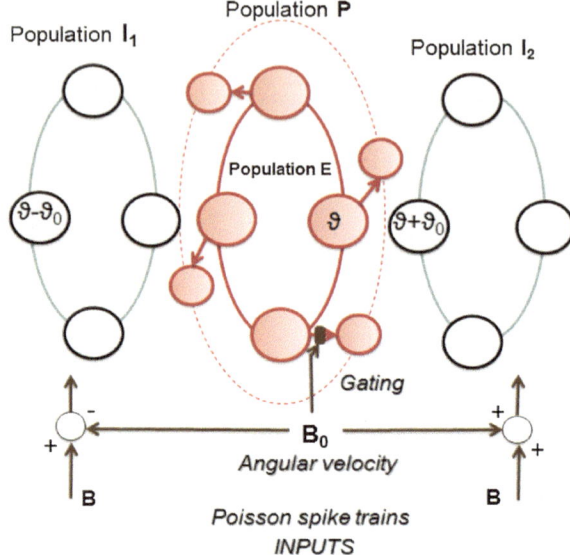

3.3 Model Simulations

The capabilities of the proposed neural structure were evaluated through a series of
simulations. The first experiments consist of positioning the hill of activity on a fixed
reference angle to initialize the network dynamics. The symmetry of the structure
determines a variable spatial distribution of the initial hill that depends on the initial
conditions. To force the formation in a given position, the landmark is applied on
a neuron coding the spatial orientation. In Fig. 3.5 the average firing rate within
populations E and I_1 is shown. The behaviour of population I_2 is similar to I_1 but
slightly shifted in order to create two boundaries on the two sides of the population
E blocking the hill of activity in a narrow area. A landmark is applied for a time of
0.5 s to the neurons next to 180° through an input current formulated as a Gaussian
function:

$$I = Ae^{-\frac{(\vartheta - \vartheta_0)^2}{2\sigma^2}} \tag{3.2}$$

where $A = 9.4\,\text{nA}$, $\sigma = 24°$ and $\vartheta_0 = 180°$.

The effect of a bias current B_0 applied for 1 s in order to unbalance the network
activity is reported in Fig. 3.6 where the shift of the hill of activity of population P is
reported.

An important difference with respect to other works [17] consists of the reduction
of synaptic connections between layers creating a more local than global network.
Moreover, the reduced number of neurons in the rings can produce residual activity
in some neurons far from the hill that in any case does not compromise the stability
of the system behaviour.

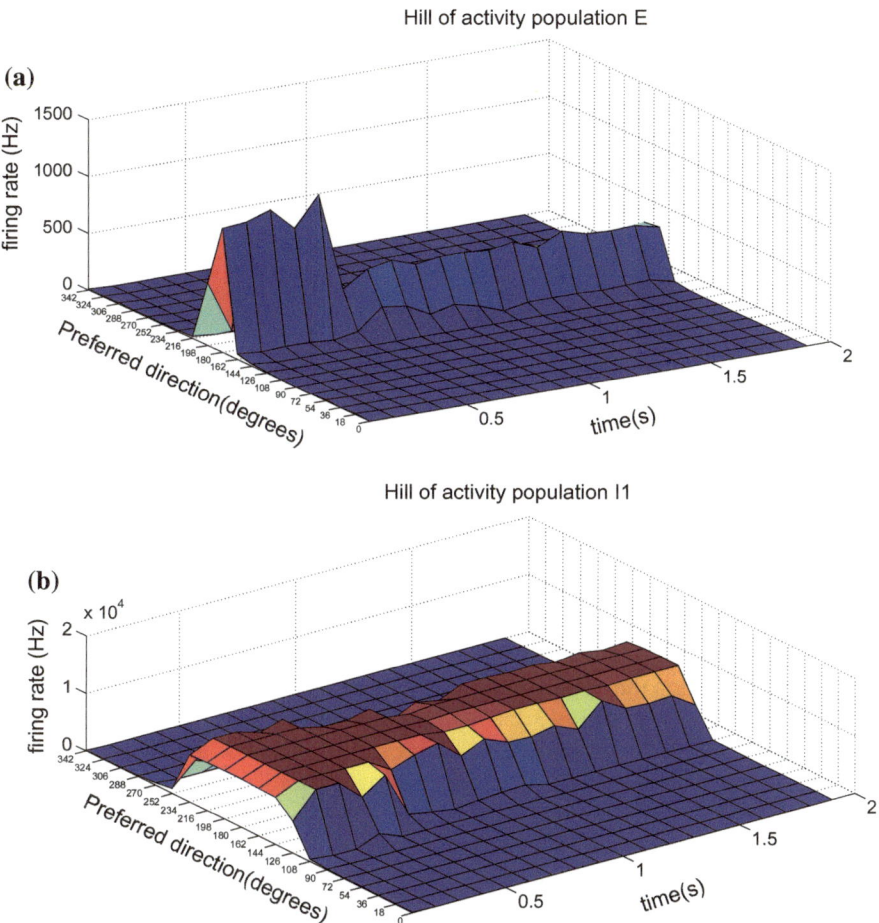

Fig. 3.5 Time evolution of the firing rate in the population E and I_1. The behaviour of I_2 is similar to I_1 with a shift on the other side with respect to population E. An external input used as landmark, is applied for 0.5 s allowing the hill formation in a specific area of the ring. This figure was reprinted from [2], © IEEE 2013, with permission

Population P integrates the activity of population E depending on the robot speed that is used as a gain (k in Eq. 3.1). A calibration procedure is used to find the gain value that converts the neural activity into a distance travelled expressed in a generic measurement unit.

To further evaluate the system performance, the network was stimulated simulating a forward movement for 1 s followed by a rotation of 130° and a second forward action for another 1 s; a constant speed of 1 m/s was considered. The angular velocity was provided to the network through a $B_0 = 920$ Hz for 0.5 s. A landmark was also provided to guide the initial formation of the hill in a position around 0°. The

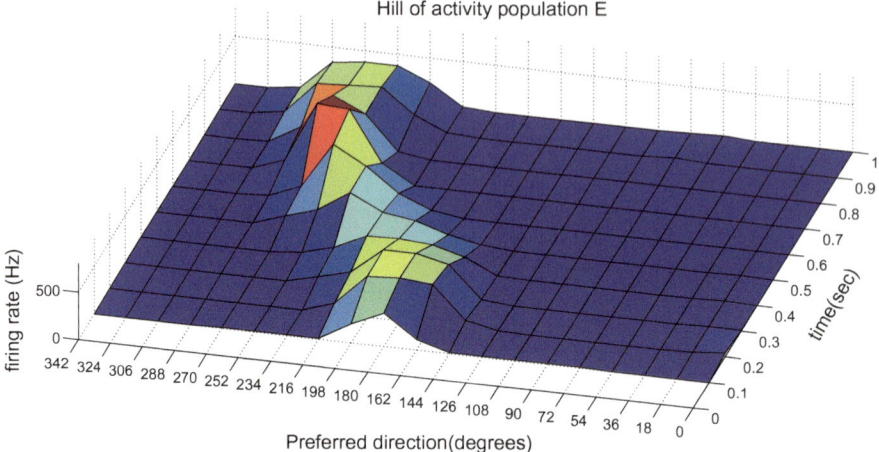

Fig. 3.6 Effect of an unbalance in the external input frequency distribution. A movement of the spiking activity of the excitatory neurons arises

network, working as a short-term spatial memory, provides as output the estimated position and orientation of the agent.

During the rotation the population P is disconnected from population E using the gating parameter $k = 0$ in Eq. 3.1 and the rotation is considered on the spot (see Eq. 3.1). The network integrates in time the speed movement of the agent making an estimation of its spatial position. The heading angle stored by the system during the 3 s simulation is reported in Fig. 3.7. The normalized activity of population P that integrates in time the state of population E, is shown in Fig. 3.8. To obtain, step by step, the internally estimated current position of the agent, a vectorial summation is performed considering neurons in population P as polar representation of spatial vectors.

A comparison between the real trajectory and the internally memorized one for a complex trajectory containing several straight lines and rotations, is reported in Fig. 3.9. The cumulated error is a consequence of the realistic processing structure and in particular it is due to the relatively low number of neurons distributed in the rings that determine the spatial resolution (i.e. about $18°$ each neuron).

Therefore, the bio-inspired neural structure can be improved for robotic applications by increasing the spatial resolution to obtain better performance. The behaviour of the system can be appreciated also in standard tests used in robotics like navigation on a square path [6]. In this simulation each rotation of $90°$ corresponds to a $B_o = 800\,\text{Hz}$ applied for 0.5 s, and each segment to a forward movement at 1 unit/s for 0.5 s. In Fig. 3.10 the real trajectory is compared with the step-by-step estimated position provided by the network.

To compare the network behaviour with the insect experiments, the detour paradigm was reproduced.

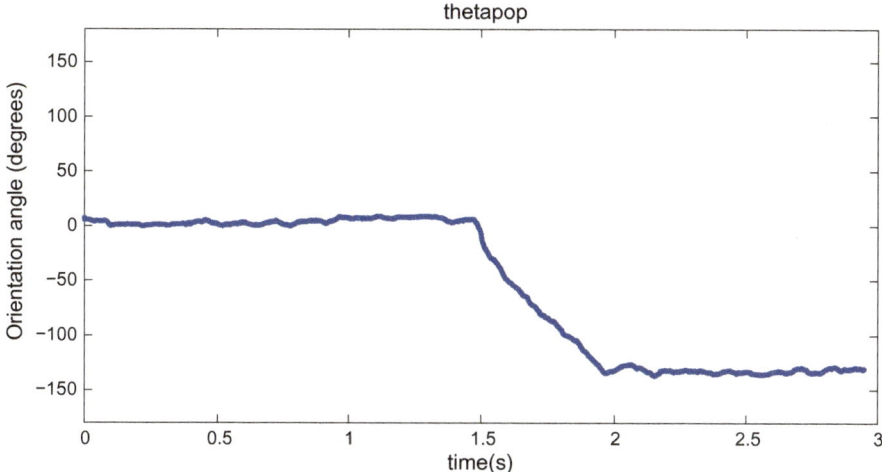

Fig. 3.7 Time evolution of the heading angle of an agent during a 3 s simulation, while performing a forward movement, a turning on the spot of 130° and a second forward movement. This figure was reprinted from [2], © IEEE 2013, with permission

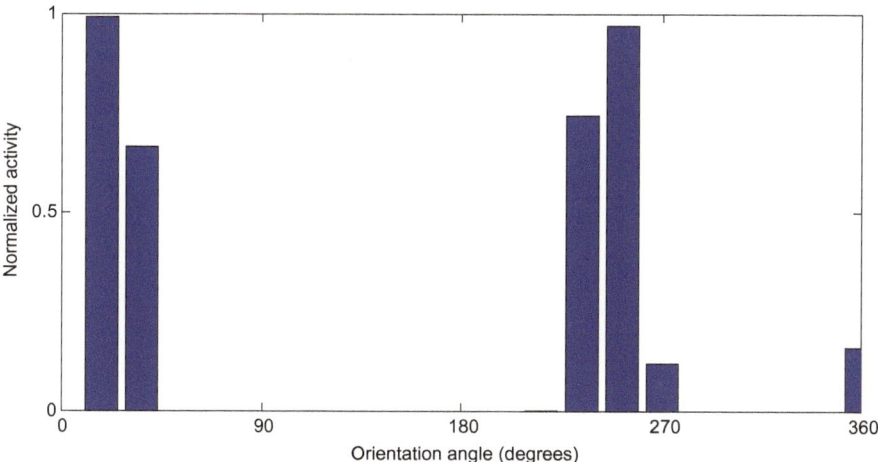

Fig. 3.8 Normalized activity of population P distributed in the different angular positions. The estimated final position of the agent is evaluated through a simple vectorial sum of the neural activities expressed in a polar coordinate system. This figure was reprinted from [2], © IEEE 2013, with permission

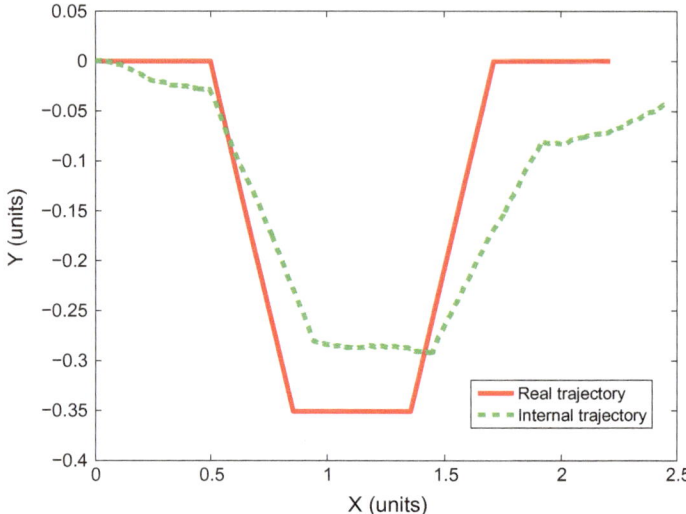

Fig. 3.9 Comparison between the real trajectory followed by the agent and the internal estimated position stored in the EB

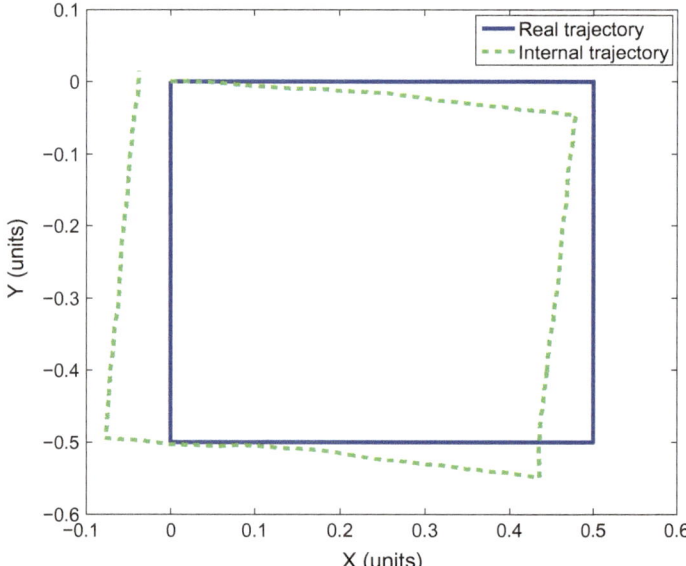

Fig. 3.10 Comparison between the real trajectory followed by the agent while performing a squared path and the estimated position stored in the EB. This figure was reprinted from [2], © IEEE 2013, with permission

The visually guided navigation in flies is mainly dependent on the central complex. When a target disappears from the scene and a new attractive object appears (i.e. a distractor), the insect stores in the EB neural structure the estimated spatial position of the obscured target that was followed in the previous time window. This estimation can be obtained using corollary discharge (efference copy) of self-motion. Distance is coded using, as measurement unit, the number of steps estimated to reach the

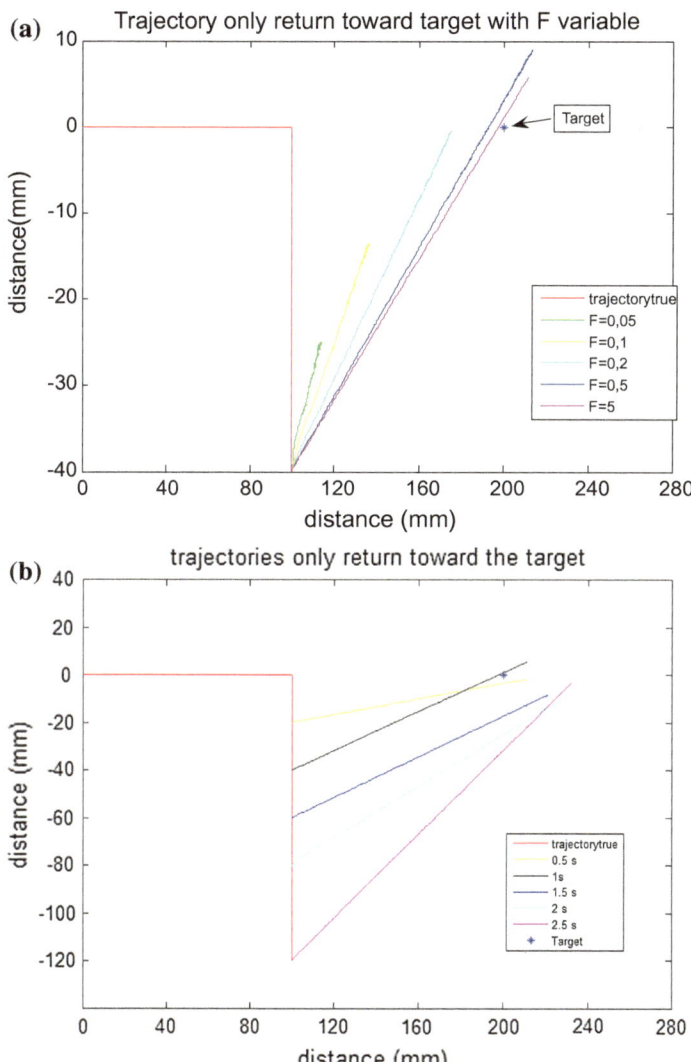

Fig. 3.11 Detour experiment: trajectory stored by the system attracted by the distractor and estimated trajectory to reach the position of the disappeared target. **a** Effect of the fading parameter F while recovering the target position. **b** The distractor was applied for different time windows from 0.5 to 2.5 s

object [3, 4]. This spatial vector is coded in the neural population P. The trajectory followed while the insect is attracted by the distractor is integrated in the network to allow the recovery of the original target position within a limited time due to the fading memory.

Three different steps can be distinguished for the simulation:

1. when the target disappears, the EB stores the target estimated position in polar coordinates charging the corresponding neuron of population P with a current proportional to the distance (coded in number of steps);
2. then the hill of activity of population E is forced to arise in a position rotated by 180°. Subsequently the path followed during the presence of a distractor placed at 90° is memorized;
3. when the distractor disappears, the information accumulated in population P is used to estimate the spatial position of the first target.

The trajectory stored by the architecture during the effect of the distractor and the estimated trajectory followed to reach the disappeared target are reported in Fig. 3.11 where the role of the fading parameter is underlined together with the effect of an increase in the distraction time.

3.4 Conclusions

In this chapter a computational model of the *Drosophila* ellipsoid body based on a four-ring spiking network is developed. The neural structure codes the heading direction using a hill of spiking activity in a ring-shaped structure. Simulation results show the capability of the system to store spatial position on a polar coordinate system. The proposed model is able to show behaviours similar to experiments with *Drosophila melanogaster* in the detour paradigm. Further details on the proposed architecture can be found in [2]; key elements contained in this model were subsequently found in the neural networks within the EB of *Drosophila* in biological experiments [16], further assessing the role of system models for a potential and fruitful subsequent biological assessment. Although the model was assessed only in simulation, the architecture proposed can be easily extended and applied on robotic roving/walking systems.

References

1. Aradi, I., Barna, G., Erdi, P.: Chaos and learning in the olfactory bulb. Int. J. Intell. Syst. **10**(1), 89–117 (1995)
2. Arena, P., Maceo, S., Patanè, L., Strauss, R.: A spiking network for spatial memory formation: towards a fly-inspired ellipsoid body model. In: The 2013 International Joint Conference on Neural Networks, IJCNN 2013, Dallas, TX, USA, pp. 1–6, 4–9 Aug 2013. https://doi.org/10.1109/IJCNN2013.6706882

3. Arena, P., Mauro, G.D., Krause, T., Patanè, L., Strauss, R.: A spiking network for body size learning inspired by the fruit fly. In: The 2013 International Joint Conference on Neural Networks, IJCNN 2013, Dallas, TX, USA, pp. 1–7, 4–9 Aug 2013. https://doi.org/10.1109/IJCNN.2013.6706883

4. Arena, P., Patanè, L., Strauss, R.: Motor learning and body size within an insect brain computational model. In: Proceedings of the Biomimetic and Biohybrid Systems—Third International Conference, Living Machines 2014, Milan, Italy, July 30–Aug 1 2014, pp. 367–369 (2014)

5. Bisch-Knaden, S., Wehner, R.: Local vectors in desert ants: context-dependent landmark learning during outbound and homebound runs. J. Comp. Physiol. **189**, 181–187 (2003)

6. Borenstein, J., Everett, H., Feng, L.: Where am I? systems and methods for mobile robot positioning. Technical Report (1996). http://www-personal.umich.edu/johannb/Papers/pos96rep.pdf

7. Collett, M., Collett, T., Srinivasan, M.: Insect navigation: measuring travel distance across ground and through air. Curr. Biol. **16**, R887–R890 (2006)

8. Cruse, H., Wehner, R.: No need for a cognitive map: decentralized memory for insect navigation. PLoS Comput. Biol. **7**(3), 1–10 (2011)

9. Haferlach, T., Wessnitzer, J., Mangan, M., Webb, B.: Evolving a neural model of insect path integration. Adapt. Behav. **15**, 273–287 (2007)

10. Hartmann, G., Wehner, R.: The ant's path integration system: a neural architecture. Biol. Cybern. **73**, 483–497 (1995)

11. Kahsai, L., Carlsson, M., Winther, A., Nassel, D.: Distribution of metabotropic receptors of serotonin, dopamine, GABA, glutamate, and short neuropeptide F in the central complex of Drosophila. Neuroscience **208**, 11–26 (2012)

12. Kuntz, S., Poeck, B., Strauss, R.: Visual working memory requires permissive and instructive NO/cGMP signaling at presynapses in the Drosophila central brain. Curr. Biol. **27**(5), 613–623 (2017)

13. Neuser, K., Triphan, T., Mronz, M., Poeck, B., Strauss, R.: Analysis of a spatial orientation memory in Drosophila. Nature **453**, 1244–1247 (2008)

14. Ofstad, T.A., Zuker, C.S., Reiser, M.B.: Visual place learning in Drosophila melanogaster. Nature **474**, 204–209 (2011)

15. Redish, A., Elga, A., Touretzky, D.: A coupled attractor model of the rodent head direction system. Netw. Comput. Neural Syst. **7**(4), 671–685 (1996)

16. Seelig, J.D., Jayaraman, V.: Neural dynamics for landmark orientation and angular path integration. Nature **521**(7551), 186–191 (2015). https://doi.org/10.1038/nature14446

17. Song, P., Wang, X.: Angular path integration by moving hill of activity: a spiking neuron model without recurrent excitation of the head-direction system. J. Neurosci. **25**(4), 1002–1014 (2005)

18. Wehner, R.: Desert ant navigation: how miniature brains solve complex tasks. J. Comp. Physiol. A. **189**, 579–588 (2003)

19. Wittmann, T., Schwegler, H.: Path integration—a network model. Biol. Cybern. **73**, 569–575 (1995)

20. Wu, C., Xia, S., Fu, T., Wang, H., Chen, Y., Leong, D., Chiang, A., Tully, T.: Specific requirement of NMDA receptors for long-term memory consolidation in Drosophila ellipsoid body. Nat. Neurosci. **10**, 1578–1586 (2007)

21. Young, J., Armstrong, J.: Structure of the adult central complex in Drosophila: organization of distinct neuronal subsets. J. Comp. Neurol. **518**, 1500–1524 (2010)

Chapter 4
Controlling and Learning Motor Functions

4.1 Introduction

Motor-skill learning is an important capability shown by animals, that need to survive in dynamically changing environments. Motor-skill learning can be considered as the process that allows the acquisition of specific movements at the aim to fulfill an assigned task. The relevance of this capability is supported by the fact that sensory-motor conditioning was one of the earliest types of associative learning discovered in insects [16].

Applying operant strategies, agents can incrementally improve their motor responses: the relevance of a movement is evaluated acquiring the available sensory signals [10, 11].

Insects, as demonstrated by several experiments, are able to perform adaptive motor-control strategies improving the joint coordination among the limbs. The gap crossing scenario is a suitable experiment to highlight motor learning capabilities, as investigated in stick insects [8, 9], *Drosophila melanogaster* [24] and cockroaches [14].

Taking into account the fruit fly experiments, exemplars with a body length of about 2.5 mm can cross gaps of up to 4.3 mm.

The analysis of the experiment using high-speed video outlined that flies perform a preliminary visual estimation of the gap width using the parallax motion generated during the approaching phase. Successively, if the gap is considered surmountable, the climbing procedure can be performed and successively improved. The learning process needs several attempts, in which a number of parameters affecting the movement coordination is modified from their values used for normal walking trying to maximize the climbing performances.

Interesting quantitative results are reported in [17] where the improvements shown by flies when they iteratively climb over gaps of the same width are illustrated.

© The Author(s) 2018
L. Patanè et al., *Nonlinear Circuits and Systems for Neuro-inspired Robot Control*,
SpringerBriefs in Nonlinear Circuits, https://doi.org/10.1007/978-3-319-73347-0_4

The presence of synaptic plasticity in MBs and its relation with motor learning, represents our working hypothesis for the development of an MB-inspired computational model [2].

Stick insects were considered as a good candidate for gap crossing experiments [8, 9]. The adaptation at the level of single leg movements was investigated, highlighting the important role of searching reflexes and coordination mechanisms. Following this analysis, specific models of gap crossing behaviour were designed as an extension of a previously developed bio-inspired locomotion control network (i.e. Walknet [12]). The obtained results were compared with biological experiments reaching a significant level of matching [8, 9]. The idea was to consider the gap crossing task as an extension of the typical normal walking behaviour with simple additional blocks needed for reflexes and coordination mechanisms. The considerations we took into account for developing motor-skill learning mechanisms are based on similar considerations, although starting from quite different perspectives. The proposed architecture is grounded on a central pattern generator for normal walking, adding plasticity only on a series of parameters to efficiently improve the climbing capabilities.

Different insects like cockroaches, served also as role models for both the simulation and implementation of obstacle climbing and gap crossing [14]. The introduction of further degrees of freedom for instance including an actuated joint in the robot body, was exploited to improve the capabilities of the robotic system to handle complex scenarios that include gap crossing and obstacle climbing [13, 14]. Other elements related to the robot kinematics and body posture, were identified to be important for solving obstacle climbing tasks, as demonstrated in [23], where a hexapod robot with a the sprawled posture was taken into account. The problem can be solved also including spoked legs that allow to improve power efficiency and walking capabilities in presence of obstacles [21]. In some cases hybrid legged and wheeled robots try to take the advantages of both solutions [4].

These previously presented approaches exploit the role of a high performing mechanical structure, an element that can be paired with other strategies where the presence of adaptive capabilities in the control mechanisms is investigated. A typical example is reported in [18]: an echo-state network was developed to generate the antenna movements in a simulated stick insect robot. The network was able to embed in the neural dynamics, specific trajectories that can be reproduced generating smooth transitions between the different attractors available in the neural reservoir.

Other interesting approaches were proposed in [13] where, using a reservoir computing architecture based on a distributed recurrent neural network, the ground contact event in each leg of a walking hexapod robot was estimated creating a forward model. The prediction error is important to improve the robot walking capabilities on different types of terrains.

The approach we are proposing can be placed in this research area which links recurrent networks and adaptive locomotion: in fact the adopted strategy takes into account the adaptive capabilities of a recurrent spiking network to develop motor learning mechanisms.

In the following sections, the MB intrinsic neurons, needed to model the motor-skill learning paradigm, are simulated using a part of the spiking network introduced in Chap. 2 that will be also adopted in Chap. 5. The structure is able to generate a rich, input-driven dynamics that is transferred to other neural centres using read-out maps that behave like MB extrinsic neurons. The learning process permits a generalization of the learned data: the network can generate the suitable output signals also when the input patterns were never presented during the learning process. Interpolating the memorized functions, the control system can generalize the adaptation process to new environmental conditions.

Another important aspect related to motor learning is the formation of a body scheme that involves a kind of adaptive calibration of an individual proprioceptive information to define a mental representation of its own body distinct from other individuals. This is achieved by detecting the correlations between the self-produced activity on the basis of the performed actions and the induced sensory information (e.g., spatial position, visual information) [25]. Taking into account the human behaviour, the construction of a peripersonal space (i.e. the reach of our hands and the step size of our legs) is performed using the visual feedback acquired from the environment in response to ones own actions.

These learning mechanisms, that for long time were thought to exist only in a few species of highly evolved and skilled animals, were recently identified also in the insect world. In particular, in *Drosophila melanogaster*, the simplest form of a body model was discovered looking at the recent experiments in which genetically identical flies can experience differences in their body-size of up to 15% due to environmental influences (e.g. food and temperature regime during the larval stages). The body-size model is acquired by each individual using visual feedback (parallax motion) from locomotion [19]. Similar experiments performed for motor learning were performed considering flies of different body-sizes. The number of unsuccessful attempts was always at maximum when the fly tried to overcome the largest just surmountable gap width. The natural consideration that emerges from these results is that flies take into account their body-size during behavioural decision processes. The body-size memory formation needs visual feedback as demonstrated with an experiment where flies hatched and lived in the dark, and could not take into account their body-size in subsequent decision making process.

On the basis of the experimental results available, a neural model is here discussed including the mechanisms used by flies to learn their body capabilities.

4.2 Working Hypotheses for the Model

To design both a biologically plausible and a computationally feasible model of the MBs, the two hypotheses were formulated:

- Different KCs accept different sensorial inputs at the level of the calyx. This assumption envisages the presence of different sensory modalities that co-exists with olfactory paths [20].
- Signal processing within the network is distributed at different levels: the locally connected neurons modeling the KCs, randomly connected with the input layer, produce spiking dynamics whereas, at the level of extrinsic neurons, we have an external learning to be applied for different tasks. This is a working hypothesis, needed to computationally simplify the model, and to implement the concept of Neural Reuse (see Chap. 6).

While designing the proposed architecture, the following structural elements were taken into account:

1. Presence of local, randomly distributed internal connections.

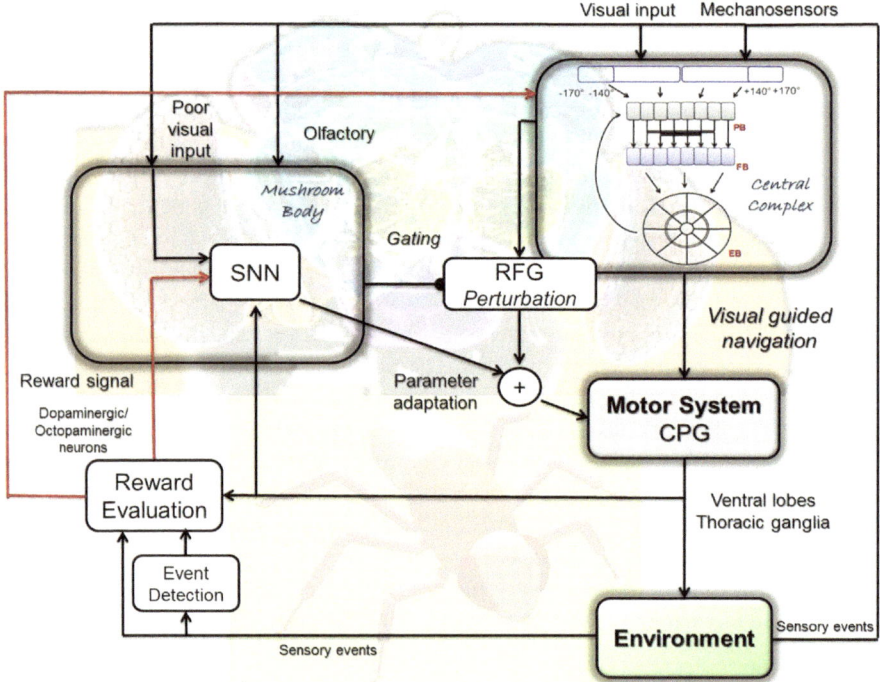

Fig. 4.1 Block diagram of the different fly brain structures involved in motor control. The parameter adaptation for the modulation of the ongoing behaviour is assumed to be performed by the MBs guided through reinforcement signals (i.e. dopaminargic/octopaminergic activity) and elaborate the learning process using a spiking neural network (SNN). The robot locomotion is guided by a Central Pattern Generator structure. A reinforcement signal is generated for the MBs whereas a Random Function Generator (RFG) is used to explore new parameters for the CPG activity. When the modulated parameters improve the final motor behaviour during the learning process, the SNN acquires the temporal evolution for the successive applications

2. Structural and functional correspondence between internal weights, mirroring the connections within the KC lattice, and the output weights, standing for connections within MBs and extrinsic neurons.
3. The same neural lattice can be concurrently used to solve completely different tasks, following the Neural Reuse paradigm, by separately training different sets of read-out maps. The unique network can therefore model a multimodal (and multifunctional) structure, in analogy to the role of MBs [7]. We are hypothesizing that different groups of extrinsic neurons are devoted to handle different tasks.

The proposed control scheme, implemented as a computational model, has been used to guide a hexapod robot simulated in a realistic dynamical environment. In relation to Fig. 4.1, the robot navigates driven by vision: the heading commands are provided to the CPG through external stimuli. An evaluation procedure assesses the suitability of the performed actions in solving the assigned task. An event detector is considered to guide the evaluation process.

The reinforcement signal is provided to the MBs to evaluate the effects of the parameter adaptation produced by the RFG that affects motor control. Successful parameter updates, leading to significant improvements in climbing behaviour are stored in a spiking neural network (SNN), considered as a plausible model for the long-term memory formation and retrieval of the best parameters selected during the learning process. Interpolation capabilities, important for the generation of feasible behaviours generalizing the situations encountered during the learning procedure, are also provided by the developed system. Finally a selector block determines when a random trial can be performed or the already learned behaviour stored in the SNN can be used for the motor activity.

4.3 Simulation Results

The proposed motor-skill learning strategy consists of continuously providing variations on leg control parameters to improve the performances in the assigned task. A smooth perturbation is applied to the parameters using as target, a part of a cosinusoidal function, considering a constrain on the final value that needs to correspond to the chosen parameter value. We considered a lattice with 8×8 neurons to model the SNN; this is a good compromise to have a considerable variety of internal dynamics available to be combined by the read-out maps. The learning process is iteratively performed in epochs to successfully memorize the needed combination of dynamics in the read-out maps. In the following simulations we considered 100 epochs with a learning rate $\eta = 0.5$. During each epoch the network is simulated for 100 integration steps. A typical activity of the neural lattice is shown in Fig. 4.2 where the presence of excitatory and inhibitory neurons is indicated.

The synaptic activity, in terms of currents generated by the lattice during learning is reported in Fig. 4.3. The read-out map creates a weighted sum of this activity to

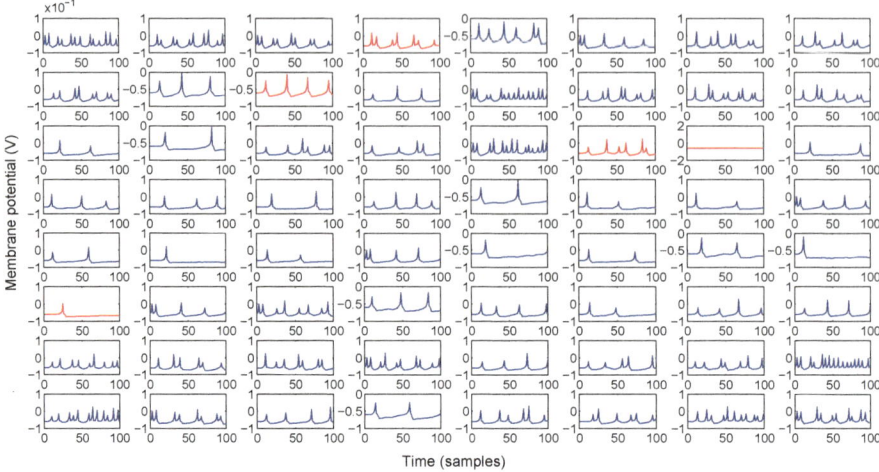

Fig. 4.2 Typical spiking activity of the lattice (i.e. SNN) while stimulated by an external input. Inhibitory neurons are outlined in red [2]

reproduce the target signals. From the figure it is evident that even a lattice with a limited number of neurons can embed a plethora of different dynamics that can be combined to obtain the needed behaviour. The presence of slightly different values for the synaptic time constant is important to determine the richness of dynamics generated by the network. The structure is extremely sensitive to a change in the input current provided to the lattice; it can generate a significant change in the temporal evolution of the network dynamics. The presence of spiking networks instead of non-spiking ones to model nonlinear dynamics is often considered as an additional complication without any tangible benefit. In the proposed structure we can demonstrate that it is indeed an added value. In fact, the results previously obtained in [3] using nonlinear non-spiking recurrent neural networks to model MB activity for solving the motor-learning problem considered a structure with 140 non-locally connected units, whereas the results here reported were obtained via a network containing only 64 spiking locally connected neurons in the liquid layer.

4.4   Motor Learning in a Climbing Scenario

On the basis of the characteristics underlined in Chap. 1, the insect brain structure can be considered as a parallel computing architecture where reflexive paths are needed for survival, whereas learned paths improve the system capabilities in generating more complex behaviours.

Analysing motor activities in insects, the thoracic ganglia are involved in the generation of locomotion gaits. The Central Pattern Generator has widely been con-

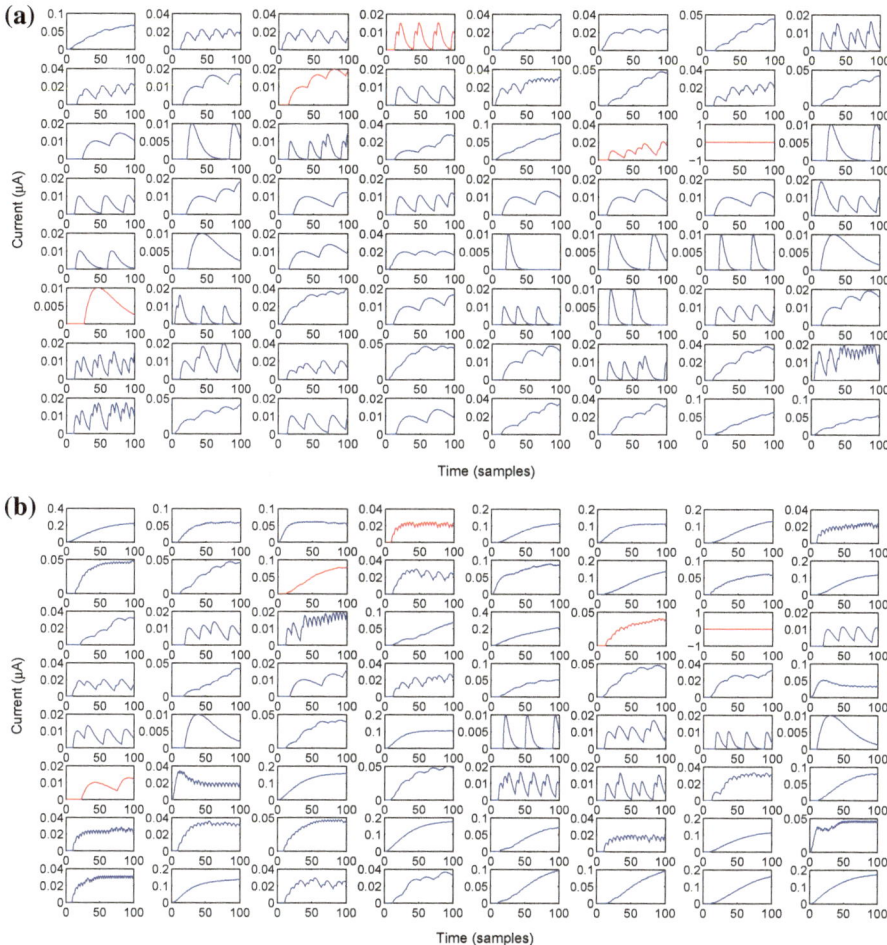

Fig. 4.3 Dynamical evolution of the activity in the output neuron when the input current is of 5μA (**a**) and 30μA (**b**). Inhibitory neurons are outlined in red [2]

sidered as a fundamental element for locomotion control where the fine-tuning of the joint movement can be achieved using sensory information. The approach proposed here considers motor learning as a strategy for modifying the basic motor trajectories on the single leg joints at the aim to improve motor-skills on the basis of the environmental constrains. Adopting a control system approach, we can model motor-skill learning through a hierarchical adaptive controller, where, in presence of different external conditions, some parameters controlling the leg joint trajectories are modulated following a trial and error-based learning process. The parameter adaptation, shaped by the kinematic constrains, realises different leg trajectories which are evaluated to assess their suitability for the on-going task. Once the initial locomotion conditions are restored, the baseline stable locomotor activity re-emerges and the acquired modulations are stored in the network to be retrieved whenever similar conditions will be encountered again.

Among the possible tasks in which the robot capabilities need to be improved, a step-climbing scenario has been considered. This simulated scenario is a possible alternative to the gap climbing scenario used in the biological experiments [9, 17, 24, 27]. Gap climbing is an affordable task for the real insect due to the adhesive capability of the fly leg tips (for the presence of pulvilli and claws), whereas this is extremely difficult for a *Drosophila*-inspired robot whose dexterity is in part limited when compared with the biological counterpart.

A step climbing task can be considered as a quite complex task for a legged robot and can be improved using an optimization method to adapt the joint movements to different surfaces. To make the system able to deal with the different aspects of the climbing process, the task was split into mutiple phases.

The first phase (i.e. approaching phase) is performed using the visual system that is able to recognize the distance between the robot and the obstacle and the obstacle height. When the robot distance from the step is below a selected threshold, the next phase is activated and the normal locomotion activity is disturbed adapting the parameters of the front legs using the Random Function Generator. The aim of this phase consists of finding a stable foot-hold on the top of the step for the anterior legs. To limit the search space, a subset of parameters available in the adopted CPG was subjected to learning in this phase. In details, we subjected only those parameters to learning which are related to front-leg movements: for the coxa joint the bias value, for the femur joint the gain value and for the tibia joint the bias and gain values. This phase is completed when there is a stable positioning of the front legs on the step, with the body lifted off. The increment of the leg joint excursion, caused by the modulation of the parameter profiles, is used as an index of the energy spent in this task and to define a reward function that is employed to optimize the adaptation process. The reward value, provided at the end of each phase, is compared with the best value reached up to this step and, if an improvement is obtained, the new sets of functions are memorized using the SNN readout map.

For the obstacle climbing task we considered an 8×8 lattice with one input (i.e. step height) and a total of ten different read-out maps, one for each parameter related to a specific leg joint and selected for the learning process. The SNN receives as input a normalized value related to the step height used to excite the lattice generating a spatio-temporal spiking activity that is integrated in a continuous, non-spiking signal, via the output synapses that converge on the output neurons, one for each readout map.

Multiple experiments were performed using an insurmountable step that need a gait adaptation to be mastered: the height of the step is around 0.9 mm, in comparison to the simulated *Drosophila* body length of 3.2 mm and the average height of the centre of mass of about 1 mm above the ground during normal walking.

The joint angular evolutions obtained through the parameter adaptation mechanisms limited to the anterior legs are reported in Fig. 4.4a. The subsequent phase is performed in a similar way: the relevant parameters to be adapted are the bias of

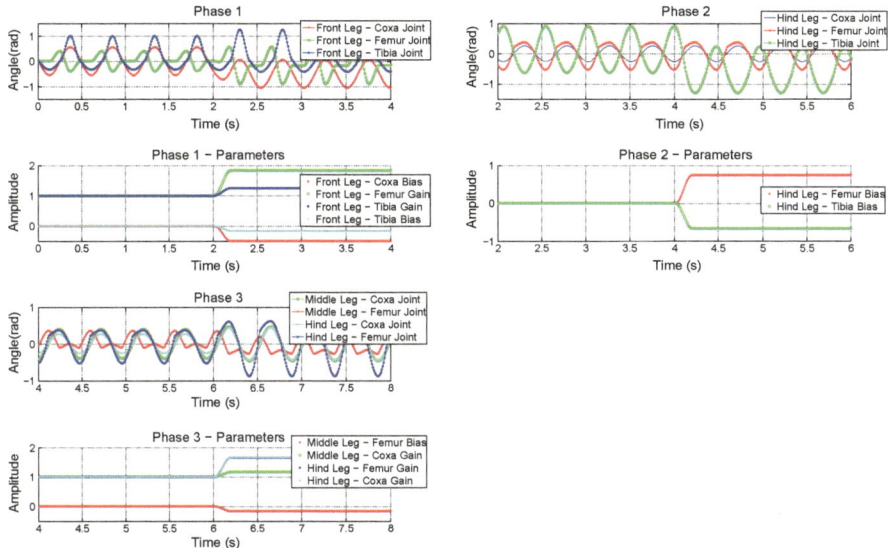

Fig. 4.4 Time evolution of the leg joint movements (only the left legs are shown) where the effect of the parameter adaptation is shown. A limited number of parameters is adapted in the three phases of the obstacle climbing task: **a** in phase 1 only the front legs are involved, **b** the hind legs in phase 2 and **c** both middle and hind legs in phase 3. The effect of the parameters on the leg joint movement is confined in the on-going phase and is absent elsewhere [2]

femur and tibia joints of the hind legs because they are important to facilitate the climbing of the middle legs. The signal used as a stopping criteria in this phase is the horizontal position of the centre of mass of the robot with respect to the obstacle. The parameter adaptations obtained during the second phase are depicted in Fig. 4.4b.

During the third phase the robot elevates the hind legs on the step: this behaviour is obtained by modulating the gain of the coxa and bias of the femur joint for the middle legs and the gain of the coxa and femur joint of the hind legs (Fig. 4.4c). In the presented experiments the function considered for the parameter modulation on the joints is a quarter of a cosinusoid. Other functions, like exponentials or sigmoids could be similarly used. The time needed by the function to reach the steady state value is a portion of a stepping cycle.

In the dynamic simulation performed, we considered an integration time $dt = 0.01\,s$ and a stepping cycle of about $1.5\,s$. Under these conditions, the parameters reach the steady state within $[20-60]$ integration steps. Analysing in details the learning progress, the RFG generates the new parameters to be tested for the first phase. If during the trial the robot reaches the success event, it is re-placed into the starting position to perform a test: this iteration is needed to guarantee the robustness of the adapted parameters. If the robot succeeds in the test, it can proceed to the second phase, otherwise the parameters are discarded and the first phase is repeated. The trial ends when the robot concludes the task or after a maximum number of

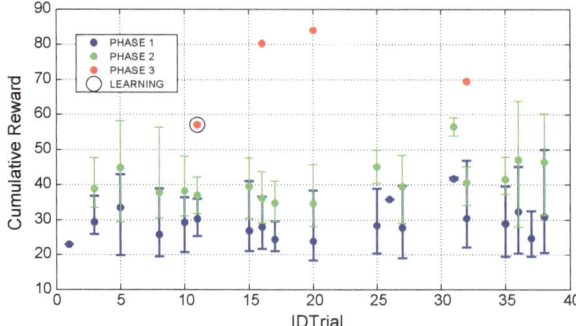

Fig. 4.5 Example of distribution of the cumulative reward for each trial; the error bars indicate the range of excursions between min and max value and the markers correspond to the mean value. The learning of the SNN is performed only for the IDTrial 11 in fact, in the success events there are no improvements for the cumulative reward in the last phase [2]

Fig. 4.6 Snapshot of the fly-inspired robot while climbing a three-steps obstacle

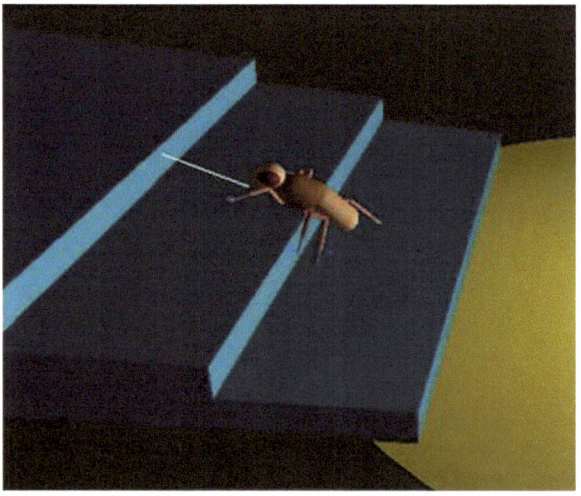

attempts (i.e. 15 events). If this time-out is reached, the parameters just used for the phases are discarded because they are not globally suitable for a complete climbing behaviour.

In Fig. 4.5 the distribution of the cumulative reward in a complete simulation is reported. For each trial the success condition for each phase can be reached multiple times until the complete climbing behaviour is tested successfully or otherwise a time-out occurs. If the cumulative reward (i.e. sum of the rewards for each phase) obtained after the last phase is lower than the best value previously reached, the parameters are learned by the network.

To evaluate the generalization capability of the control system, the previously learned system was evaluated in a different scenario where a stair-like obstacle

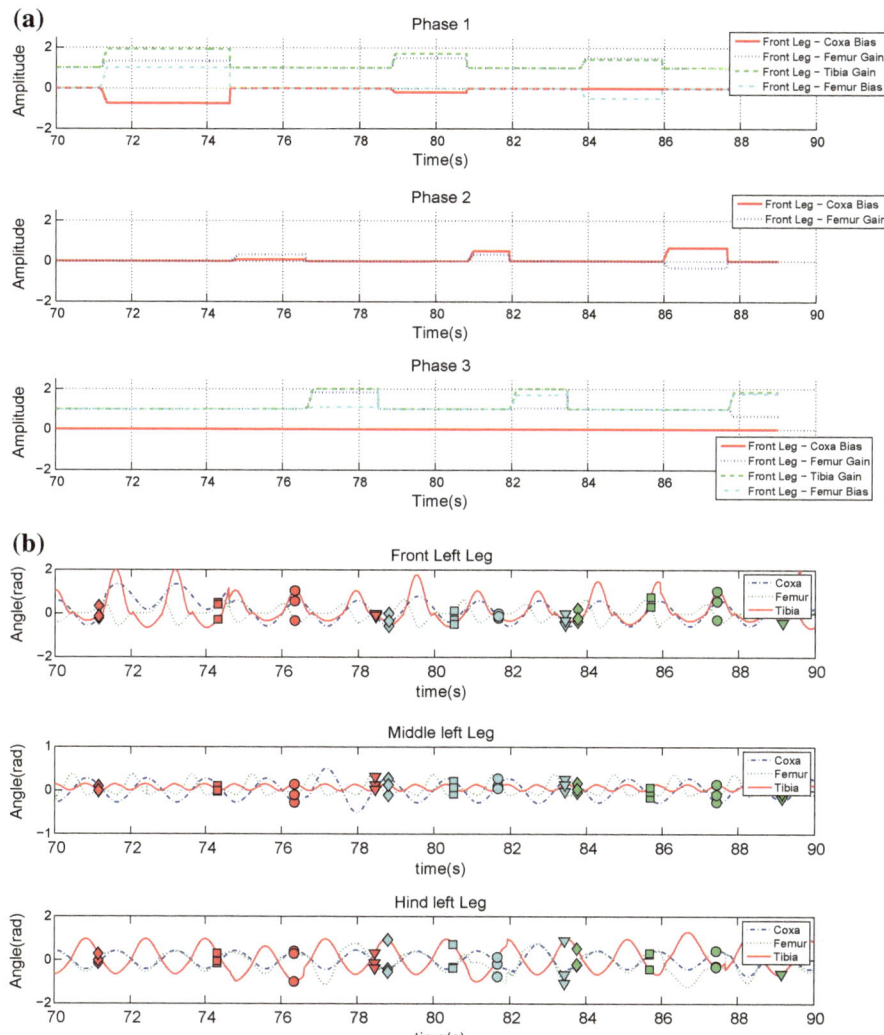

Fig. 4.7 **a** Time evolution of the parameters used during the three different phases of the climbing procedure. **b** Trend of the joint positions for the left side legs when the robot faces a series of obstacles

was introduced (see Fig. 4.6). The robot followed the same climbing procedure as described above, repetitively applied for the three stair steps encountered on its path with height 1.3 mm, 1.1 mm and 0.9 mm, respectively. The detection of each obstacle produces an effect at the motor level on the basis of the parameter adaptation mechanism generated by the SNN. Figure 4.7a shows the trend of the joint position angles for the left-side legs during the complete climbing task and the parameters adopted during the different phases (Fig. 4.7b). The adapted parameters produce changes in

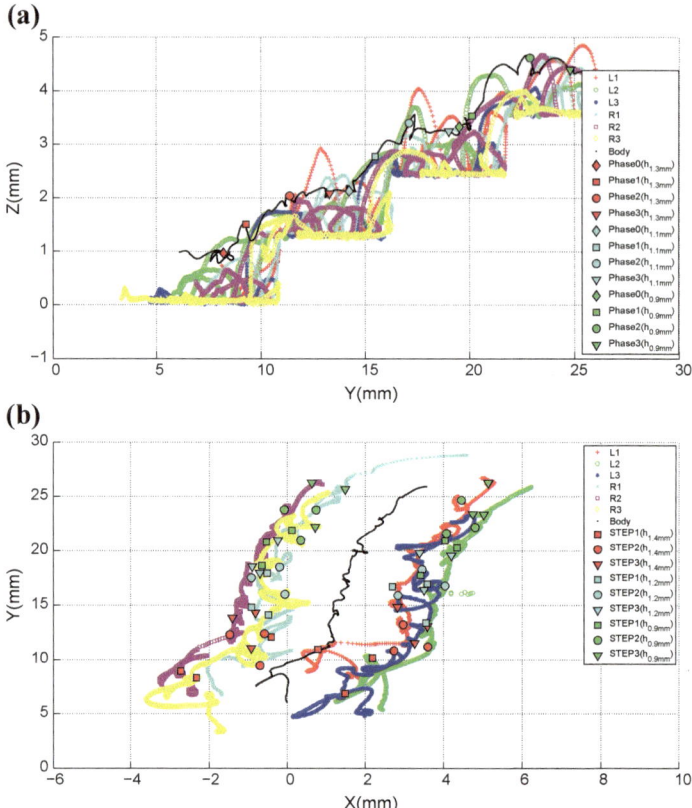

Fig. 4.8 b Trajectories followed by the centre of mass of the robot and by the tip of each leg during the climbing behaviour facing with multiple obstacles with height 1.3 mm, 1.1 mm and 0.9 mm, respectively. Markers are placed in the signals to indicate when the robot completes each climbing phase. Trajectories along the X-Y plane of the relevant elements of the robot (e.g. COM, leg tips) followed during the climbing procedure of the three-steps obstacle [2]

the leg movements during the different climbing phases as illustrated in Fig. 4.8 where the evolution of the robot centre of mass and the leg tip positions are reported.

The results are based on the adaptive capabilities of the legs acquired during the motor-skill learning phase. The degrees of freedom of the robot are located only on the legs and the body structure was considered mostly rigid as in the fruit fly case. Including in the robotic structure active body joints [13], mimicking the body of other insects like cockroaches, would further improve the robot capabilities. Therefore, the proposed control strategy can be also applied to other and different robotic structures to improve their motor capabilities in fulfilling either obstacle climbing tasks or other similar scenarios affordable for the robot under consideration.

To augment the knowledge of the system to improve the adaptation process, other information on the robot structure can be included; for instance the posture

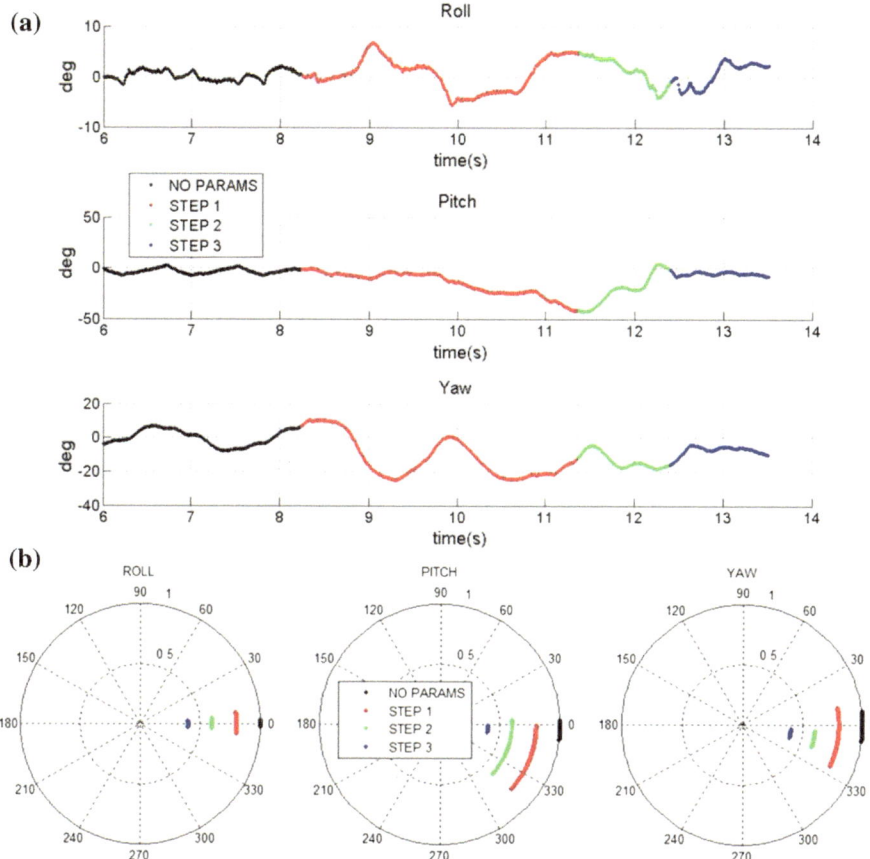

Fig. 4.9 a Time evolution of the robot attitude during the climbing procedure of a single step. **b** Distribution of roll, pitch and yaw depending on the climbing phase

of the system during the climbing procedure is dependent on the current phase of the climbing behaviour. Figure 4.9 shows the time evolution of the roll, pitch and yaw variables and the angular distribution during the different phases. These data could be relevant for the parameter choice and will be taken into account in future developments.

4.5 Body-Size Model

The multisensory representation of our body plays a crucial role in the development of many higher-level cognitive functions [22].

As robotic systems become more complex and versatile, there is a growing demand for techniques devoted to automatically learn the body schema of a robot with only minimal requirement for human intervention [26]. We can consider the body model as a sensory-motor representation of the body used to fulfill specific behaviours; it would contain both *short-term* (e.g. position of a limb at a given instant) and *long-term* dynamics (e.g. biomechanical characteristics and dimension of limbs) [15].

The mental processes that guide the formation of a body model, involve an adaptive calibration of individual proprioceptive fields to define a task-oriented representation of its own body. This is achieved by detecting the causal effects between the self-generated activities and the corresponding sensory information. In *Drosophila melanogaster*, the simplest form of a body model is envisaged. Visual feedback through parallax motion from locomotion is required to acquire and calibrate the memory for the body-size acquisition [19]. In gap climbing experiments with flies of different body-sizes, the number of unsuccessful attempts was always maximum at the largest just surmountable gap width. This result demonstrates that flies take into account their body-size during the decision-making process.

The role of CX in modelling body-size knowledge formation is fundamental. A plausible model of CX includes a distributed structure spatially associated with the system's visual field able to acquire spatial information in terms of angular position between the objects of interest and the robot. A plausible model of the CX is shown in Fig. 4.10 where the different elements identified in this neural assembly are reported. The visual system acquires information on relevant objects in the scene, transferring the *where* and *what* to the protocerebral bridge (PB) and the fan-shaped body (FB) that are responsible for heading control and visual learning, whereas the ellipsoid body (EB) is involved in spatial memory formation.

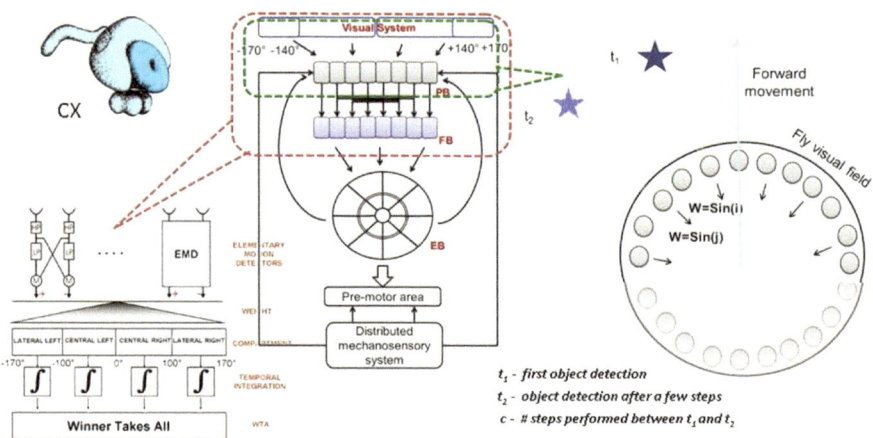

Fig. 4.10 Block scheme showing the different elements identified in modelling the CX. The visual system transfers information to the PB structure involved in body-size learning. The FB participates in visual learning and orientation control, whereas the EB is in charge for spatial memory formation

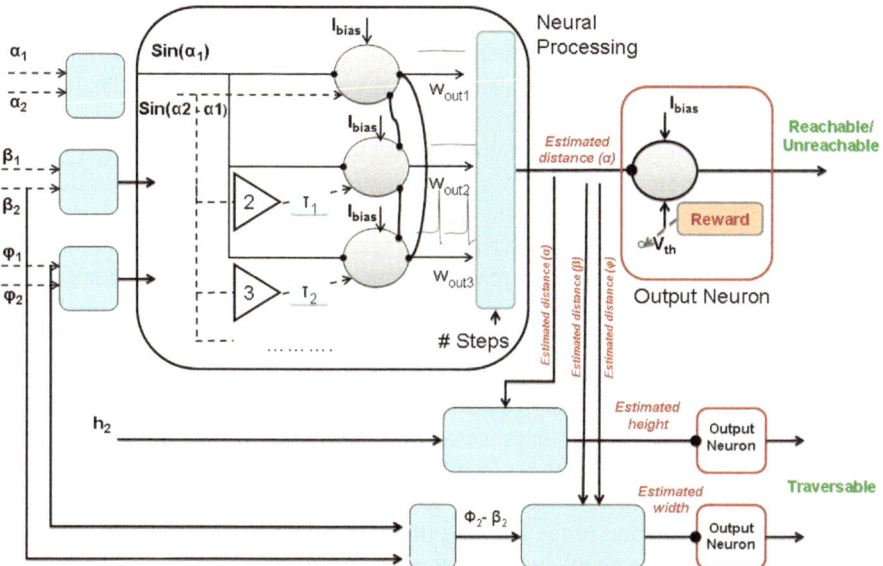

Fig. 4.11 Block scheme of the control structure devoted to learn relations between the robot body-size and the objects in the environment. The robot generates parallax motion moving on a straight line for a fixed number of steps. The view angle between the relevant points identified in the detected objects are acquired through the visual system as shown in Fig. 4.12 where a door is taken into account. By a series of processing blocks, the model is able to estimate the distance from the object in steps and, considering the door case, also the height and the width. A series of output neurons fire, if the object is reachable (i.e. its estimated distance is below a threshold) and if, in the case of a door, the robot can be pass through. The learning mechanisms are based on a reward signal provided to the robot depending on the results obtained while approaching the object. A series of input currents (i.e. $g_a V_t h$ where $g_a = 1$) are consequently tuned for the calibration of the system's body-size model

The designed model already tested for direction control, spatial memory and other capabilities [1, 5], was extended to include body-size knowledge formation. A scheme of the neural processing is reported in Fig. 4.11.

The spatial orientation of the objects with respect to the robot is acquired in two different time instants: before and after a forward movement action for a fixed number of robot steps (e.g. 5 steps). The acquired orientations of the point of interest in the selected object (e.g. centre of mass, vertex and others) are processed through a constant weight that is related to the sinusoidal function of the corresponding angle. The signals are then processed through a series of spiking neurons connected in a winner-takes-all topology. One of the inputs inhibits all neurons and the other excites each neuron with different weights (i.e. increasing weight and time delay). The first-firing neuron wins the competition and its weighted output, modulated by the number of performed steps, represents an estimation of the distance from the object of interest to the actual robot position. A final stage includes an output neuron that works as a gate to determine if an object is considered reachable or not depending on its estimated distance. The correct decision is learned through a reward-based process that guides an adaptive threshold mechanism [6]. The model has been extended to estimate not only the distance but also the height and the width of objects that, for instance, could be candidate gates to new areas to be explored. A correct body-size model will allow to choose the best door to pass through.

Using the experimental results available, a neural model is here proposed, based on the mechanisms used by flies to learn their body capabilities. The architecture has been applied to a hexapod simulated robot in a scenario where the robot should learn through a reward-based system the reachable/unreachable space in the arena within a fixed step number and the presence of potential doors to pass through, suitable for the robot dimensions.

During the parallax evaluation procedure the information on the relevant angles between the robot and the object (i.e. a door) of interest are acquired as shown in Fig. 4.12 and used in the decision process.

To test the model, the robot was placed in an arena consisting of four rooms, each one including three doors with different dimensions. The robot, placed in one of the rooms at the beginning of the learning phase, through parallax motion evaluates the relevant information about each passage (i.e. the estimated distance, height and width) and tries to pass through one randomly chosen door. Depending on the success

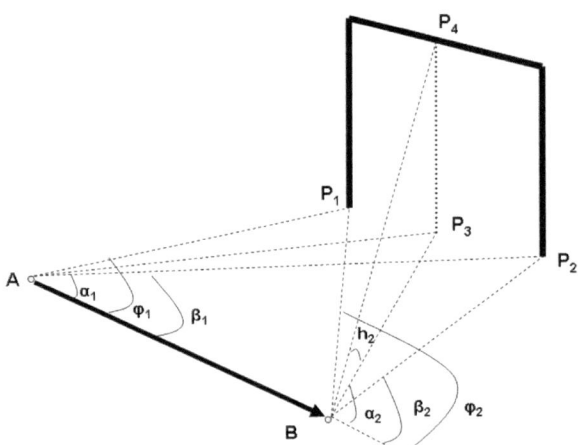

Fig. 4.12 The robot, starting from position A, moves straightforward reaching position B in a fixed number of steps. To provide the needed input to the computational model, a series of angles are acquired from the vision system. The object taken into account is a door and the points of interest are the two extreme positions on the ground P_1 and P_2 and the middle point P_3 that is the projection of the point P_4 on the ground. The α angles are considered between the line $\overline{AB}$ and the point of interest P_3; the β angles are related to the point of interest P_2 whereas the ϕ angles are related to the point of interest P_1. Finally the h_2 angle represents the elevation of point P_4 when the robot is in B

Fig. 4.13 Trajectory performed by the robot while exploring the environment during the learning phase used to acquire the needed knowledge for the body-size model formation

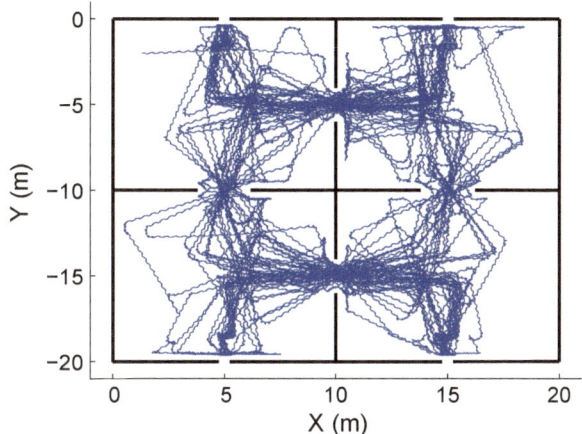

of the performed action, a reward signal is triggered and used to adapt the output neuron threshold accordingly, for the tuning of the body-size internal model. The traversable passages are placed between two adjacent rooms whereas all the other doors are too small to be used. This arena configuration allows to confine the robot to the four-rooms structure for continuous autonomous learning.

The trajectory followed by the robot while exploring the environment is shown in Fig. 4.13.

During the learning phase, the robot performs a forward motion of five steps needed to acquire parallax information on the objects visible in the scene (i.e. the three doors present in each room). Thereafter, one object is randomly selected and the robot orients towards it. During the successive approach, the number of travelled steps is acquired and used to trigger a reward event that indicates whether the selected object was within the reachability area that was fixed to 12 steps. When the robot is next to the door, the actual heading with respect to the door centre is evaluated and if needed corrected. A procedure is then activated to evaluate whether the door is traversable, otherwise the robot performs an escaping behaviour moving backward and reorienting towards the room centre. If a success occurs, the neuron thresholds are updated and the procedure is repeated [6].

When the output neuron threshold converges to steady state values, a testing phase can be performed. The robot is now able to evaluate which door is reachable and passable avoiding to be stuck in trying to pass through too narrow passages. The trajectory obtained during the testing phase is shown in Fig. 4.14; the attempts to pass through the door placed on the external walls are no more present.

Fig. 4.14 Trajectory
followed by the robot during
the testing phase. The
body-size model is used to
chose the suitable doors to
pass through avoiding the
others

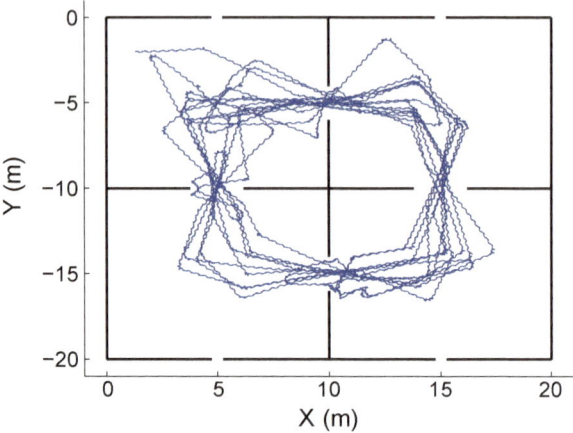

4.6 Conclusions

In this chapter a bio-inspired, embodied, closed-loop neural controller has been
designed and implemented in a simulated hexapod robot that is requested to improve
its motor-skills to face unknown environments. Taking inspiration from the insect
brain and in particular from the fruit fly, the following hypotheses were formulated:
relevant role of MB neuropiles in the motor learning task; direct transfer of the
important role of transient dynamics in the olfactory learning from the locust to the fly
brain and further extension to motor learning; design of a neuro-computational model
based on a liquid state machine-like structure for the implementation of obstacle
climbing in a simulated hexapod robot.

In details, a computational model for motor-skill learning was developed and
realized in a dynamic simulation environment. Inspired by behavioural experimental
campaigns of motor learning in real insects, the computational structure consisted of
a randomly connected spiking neural network that generates a multitude of nonlinear
responses after the presentation of time dependent input signals. By linearly com-
bining the output from the lattice neurons with a weighted function, a reward-based
strategy allows to learn the desired target by tuning the weights of a readout map.
The robot was also able to deal with step heights never presented before, exploiting
the interpolation abilities of the proposed network. The body-size model was also
taken into account, light was shed on the underlying neural mechanisms and they
were integrated into the insect brain architecture.

References

1. Alba, L., Morillas, S., Listan, J., Jimenez, A., Arena, P., Patanè, L.: Embedding the anafocus eyeris vision system in roving robots to enhance the action-oriented perception. In: Proceedings of Microtechnologies for the New Millennium (SPIE 09), pp. 7365–08. Dresden, Germany (2009)
2. Arena, E., Arena, P., Strauss, R., Patanè, L.: Motor-skill learning in an insect inspired neuro-computational control system. Front. Neurorobot. **11**, 12 (2017). https://doi.org/10.3389/fnbot.2017.00012
3. Arena, P., Caccamo, S., Patanè, L., Strauss, R.: A computational model for motor learning in insects. In: International Joint Conference on Neural Networks (IJCNN), Dallas, TX, pp. 1349–1356, 4–9 August 2013
4. Arena, P., De Fiore, S., Patanè, L., Pollino, M., Ventura, C.: Insect inspired unsupervised learning for tactic and phobic behavior enhancement in a hybrid robot. In: WCCI 2010 IEEE World Congress on Computational Intelligence, pp. 2417–2424. Barcelona, Spain (2010)
5. Arena, P., Maceo, S., Patanè, L., Strauss, R.: A spiking network for spatial memory formation: towards a fly-inspired ellipsoid body model. In: IJCNN, Dallas, TX, pp. 1245–1250 (2013)
6. Arena, P., Mauro, G.D., Krause, T., Patanè, L., Strauss, R.: A spiking network for body size learning inspired by the fruit fly. In: Proceedings of International Joint Conference on Neural Networks, Dallas, TX, pp. 1251–1257 (2013)
7. Arena, P., Patanè, L., Strauss, R.: The insect mushroom bodies: a paradigm of neural reuse. In: ECAL, pp. 765–772. MIT Press, Taormina, Italy (2013)
8. Bläsing, B.: Crossing large gaps: a simulation study of stick insect behavior. Adapt. Behav. **14**(3), 265–285 (2006)
9. Bläsing, B., Cruse, H.: Mechanisms of stick insect locomotion in a gap crossing paradigm. J. Comp. Physiol. **190**(3), 173–183 (2004)
10. Brembs, B., Heisenberg, M.: The operant and the classical in conditioned orientation of Drosophila melanogaster at the flight simulator. Learn. Mem. **7**(2), 104–115 (2000)
11. Broussard, D., Karrardjian, C.: Learning in a simple motor system. Learn. Mem. **11**, 127–136 (2004)
12. Cruse, H., Kindermann, T., Schumm, M., Dean, J., Schmitz, J.: Walknet a biologically inspired network to control six-legged walking. Neural Netw. **11**, 1435–1447 (1998)
13. Dasgupta, S., Goldschmidt, D., Wörgötter, F., Manoonpong, P.: Distributed recurrent neural forward models with synaptic adaptation and cpg-based control for complex behaviors of walking robots. Front. Neurorobot. **9**, 1–10 (2015). https://doi.org/10.3389/fnbot.2015.00010
14. Goldschmidt, D., Wörgötter, F., Manoonpong, P.: Biologically-inspired adaptive obstacle negotiation behavior of hexapod robots. Front. Neurorobot. **8**, 3 (2014). https://doi.org/10.3389/fnbot.2014.00003
15. Hoffmann, M., Marques, H., Hernandez, A., Sumioka, H., Lungarella, M., Pfeifer, R.: Body schema in robotics: a review. Auton. Ment. Dev **2**(4), 304–324 (2010)
16. Horridge, G.: Learning of leg position by headless insects. Nature **193**, 697–8 (1962)
17. Kienitz, B.: Motorisches lernen in Drosophila Melanogaster. Ph.D. thesis (2010). Shaker Verlag, Aachen
18. Krause, A., Bläsing, B., Dürr, V., Schack, T.: Direct control of an active tactile sensor using echo state networks. In: Ritter, H., Sagerer, G., Dillmann, R., Buss, M. (eds.) Human Centered Robot Systems. Springer (2009)
19. Krause, T., Strauss, R.: Mapping the individual body-size representation underlying climbing control in the brain of *Drosophila melanogaster*. In: 33rd Gottingen Meeting of the German Neuroscience Society, pp. T25–11B (2011)
20. Lin, A., Bygrave, A., de Calignon, A., Lee, T., Miesenböck, G.: Sparse, decorrelated odor coding in the mushroom body enhances learned odor discrimination. Nat. Neurosci. **17**, 559–568 (2013)

21. Moore, E., Campbell, D., Grimminger, F., Buehler, M.: Reliable stair climbing in the simple hexapod RHex. In: IEEE International Conference on Robotics and Automation, pp. 2222–2227 (2002)

22. Nabeshima, C., Lungarella, M., Kuniyoshi, Y.: Timing-based model of body schema adaptation and its role in perception and tool use: A robot case study. In: 4th IEEE International Conference on Development and Learning (ICDL-05), Osaka, Japan, pp. 7–12 (2005)

23. Pavone, M., Arena, P., Fortuna, L., Frasca, M., Patanè, L.: Climbing obstacle in bio-robots via CNN and adaptive attitude control. Int.J. Circuit Theory Appl. **34**, 109–125 (2006)

24. Pick, S., Strauss, R.: Goal-driven behavioral adaptations in gap-climbing Drosophila. Curr. Biol. **15**, 1473–8 (2005)

25. Pitti, A., Mori, H., Kouzuma, S., Kuniyoshi, Y.: Contingency perception and agency measure in visuo-motor spiking neural networks. IEEE Trans. Auton. Ment. Dev. **1**(1) (2009)

26. Sturm, J., Plagemann, C., Burgard, W.: Adaptive body scheme models for robust robotic manipulation. In: RSS—Robotics Science and Systems IV (2008)

27. Triphan, T., Poeck, B., Neuser, K., Strauss, R.: Visual targeting of motor actions in climbing Drosophila. Curr. Biol. **20**(7), 663–668 (2010)

Chapter 5
Learning Spatio-Temporal Behavioural Sequences

5.1 Introduction

Animal brains can be studied by modelling relevant neural structures on the basis of behavioural experiments. This research continuously improves our knowledge about learning mechanisms. The developed architectures have been deeply investigated in the last decades both to understand the sources of the impressive animal capabilities and to design autonomous systems and control strategies able to reach improved levels of autonomy in robot acting in non-structured environments.

A deep analysis of the sequence learning processes existing in living beings is a hard task, however insects can represent an interesting starting point. In fact, in insects, neurobiological evidence is provided for processes that are related to spatio-temporal pattern formation and also learning mechanisms that can be used to solve complex tasks including also sequence learning.

As deeply discussed in Chap. 1, there are different types of olfactory receptor neurons found in *Drosophila melanogaster*, whose collective dynamics contributes to the encoding of the features (e.g. odorant components) of the source providing the stimuli. The antennal lobes (ALs) are the first neuropil in the olfactory path; they consist of glomeruli linked to olfactory receptor neurons coming from the antennal receptors. Information is passed on to projection neurons (PNs), which project to protocerebral areas [42]. The connection with the large number of MB cells allows a boost in dimensionality useful to improve the representation space [12, 46]. At the same time PNs are connected to the lateral horn (LH). In locusts LH inhibits, after a delay, the activity of the Kenyon cells (KCs) neurons [34]. Therefore the KCs receive a sequence of excitatory and inhibitory waves from the PNs and LH, respectively and are believed to communicate with one-another through axo-axonal synapses [17]. In this chapter as well as in the previous one, we modelled the KC layer as a dynamic spatio-temporal pattern generator extracting the relevant dynamics needed to perform a behavioural task by training multiple read-out maps.

© The Author(s) 2018
L. Patanè et al., *Nonlinear Circuits and Systems for Neuro-inspired Robot Control*,
SpringerBriefs in Nonlinear Circuits, https://doi.org/10.1007/978-3-319-73347-0_5

To understand neural circuits in insect brains, several approaches can be taken into account and merged: both behavioural and neurophysiological experiments, and the realization of computational models at different levels of complexity. Several different examples of MB models were recently proposed and are available in the literature. One of the first MB models developed for olfactory associative learning was introduced by [15]. This model is focused on analysing olfactory conditioning and the effect of positive and negative reinforcement signals. On the basis of the experimental biological evidences, other ideas were exploited to design biologically plausible models of the MBs' neural activity and behavioural functionalities [41].

The self-organization properties available in the MBs are fundamental when the sequence learning problem is considered. An interesting analysis of this aspect was presented in [33], where a model based on spiking neurons and synaptic plasticity, distributed through different interacting layers was proposed. In their studies the MBs are assumed to be multi-modal integration centres involving both olfactory and visual inputs. As strongly supported in our model, their system capabilities are independent of the type of information processed in the MBs.

Other works investigated the interaction of MBs and ALs in non-elemental learning processes [49]. Different levels of learning including reinforcement mechanisms were adopted at the level of the KCs with the aim to develop a non-elemental learning strategy. Our proposed architecture considers learning at the KC layer and also plasticity at the level of the AL as suggested by [38] where filtering mechanisms were applied to reduce noise and reconstruct missing features at the beginning of the neural process.

The role of time is also important as emphasised in the locust olfactory processing [48]. To encode complex natural stimuli such as odours we need to consider the precise timing of the neural activity. All these aspects were considered in our model, where the olfactory system and neural circuits are modelled using dynamical systems able to generate different neural activities that can be associated to a series of behaviours. The time evolution is mapped into a space-distributed dynamic which can be adapted to generate a multitude of concurrent behaviours, among which are sequence learning and retrieval.

The problem of modelling biological nervous system functions by neural dynamics was actively investigated [1, 2]. Different kinds of spiking-based networks were taken into account for the development of the proposed architecture [4, 23].

The designed model was developed in different stages that allowed to continuously improve the functionalities included. In the first neural structure that will be presented, the KCs dynamics converges onto a cluster of activity directly related to the input [3, 5]. The network topology, chosen to obtain this behaviour, is inspired by the winner-takes-all solution: it includes in the lattice local excitatory and global inhibitory connections. Successively, a second approach will be also discussed where we decided to model the KCs' activity as a Liquid State Network (LSN), a lattice of connected spiking neurons similar to a Liquid State Machine [28], that contains mainly local synaptic connections (as in a Cellular Nonlinear Network structure [3, 7] already used for locomotion control [8]), resembling axo-axonal communication among the KCs [17]. This lattice modulates sensory information, creating a dynamic map, which can be exploited concurrently both for classification and for motor learning purposes. Taking into account the results obtained in other works [21], the classification task was developed using the sparse dynamics generated within the KC lattice of neurons implemented using a LSN, where an equilibrium in the firing rate is not requested and the neural activity can continuously change in time. In this scenario, multiple read-out maps can exploit this far-from-equilibrium neural activity (as proposed in [36]) to extract the suitable dynamics needed to solve the on-going task.

In the literature there are different MB-inspired models for classification with structures mainly based on several lattices of spiking neurons [32, 40]. In the here presented model we included a new layer, named context layer, needed to develop sequence learning capabilities into the architecture.

To summarize the main capabilities of the developed system, the relevant behaviours that can arise from this unique model are reported in Table 5.1.

Table 5.1 Different behaviours that can be obtained using the proposed architecture. For each behaviour the involved neural structures together with the relevant learning aspects are reported

Behaviours	Neural structures involved	Plasticity
Attention	Antennal Lobe (AL)	(1) STDP from α-β-lobes to AL
	α-β-lobes	(2) Memory effect in the α-β-lobes
Delayed match-to-sample	Antennal Lobe (AL)	(1) STDP from α-β-lobes to AL
	α-β-lobes	(2) Feedback synapses from α-β-lobes
	α'-β'-lobes	to α'-β'-lobes and vice-versa
	Sameness neuron	(3) Activity detection by the Sameness Neuron
Expectation	Antennal Lobe (AL)	(1) STDP between one feature to
	α-β-lobes	other features within the AL
	Context layer	(2) STDP From α-β-lobes to AL
		(3) STDP from contex layer to α-β-lobes
Sequence learning	Antennal Lobe (AL)	(1) STDP From α-β-lobes to AL
	α-β-lobes	(2) STDP from context layer to α-β-lobes
	Contex layer	(3) STDP from context layer to output layer
	Output layer	(4) STDP from α-β-lobes to output layer
Motor learning	Central complex (CX)	(1) Gating function between CX and MBs
	Intrinsic and extrinsic KC	(2) Read-out maps learning
	Output layer	

Going deeper into details, starting from the basic capabilities of the system, the persistence/distraction mechanisms can be presented. Wild-type insects can follow a target and thereby avoid fickle behaviour that can arise when distracters are introduced in the scene. Flies with inactivated MBs lose this capability as demonstrated in different experiments with *Drosophila melanogaster*: the attention is continuously switched from the target of interest and the distracter with a considerable worsening in terms of time and energy spent in the process. In the proposed model this attentional capability is performed using feedback connections which produce a memory effect at the level of the KCs in the α-β-lobes. When, in analogy to the MB defective flies, such links are suppressed, the loss of attention is obtained.

Another important capability available in the proposed model is a solution for the delayed match-to-sample task. As illustrated in details in [11], the introduction of the α'-β'-lobes in the architecture allows to identify the presence of two successive presentations of the same element through the detection of an increment in frequency in the α-β-lobes' activity. The acquired information can be also used to elicit, after conditioning, a specific behaviour that can be triggered by a matching detection.

The potentialities of the developed MB-inspired architecture are increased with the introduction of a layer that could be related to the γ-lobe, here called Context layer. This neural structure is used to store information about the sequence of events previously acquired by the system. This capability is relevant for evaluating the neurally encoded causality between consecutively presented objects; expectations on the successive presentation can emerge from this structure [3]. The presence of context layer improves the expectation performance that can be extended from one-step predictions to reproduce sequences of objects, solving also potential ambiguities, exploiting the context that is behind each object.

5.2 Model Structure

A first scheme of the proposed architecture is reported in Fig. 5.1. The connections' shape and the weight distribution allow the network to create clusters of activities as shown in Fig. 5.2 where the formation of a cluster of neural activity in the $\alpha - \beta$-lobe neurons is shown in different time windows.

5.2.1 Antennal Lobe Model

Inspired by the insects' ALs, the input layer is able to codify either the odour components (i.e. odorants) or, in a more general scenario, the extracted *features* of presented objects. In the insect ALs each glomerulus receives input from just one type of olfactory receptor; in our model each neuron in the input layer encodes a particular feature related to the object of interest. The AL model contains several neurons organized in groups used to codify a type of feature. The pool of neurons in each group

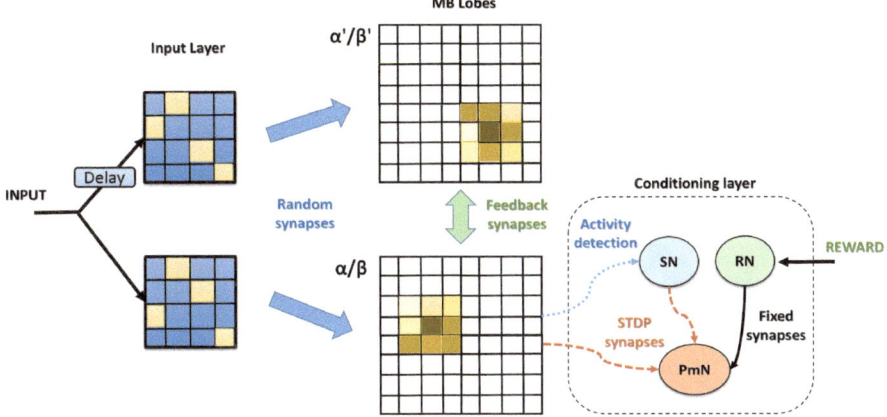

Fig. 5.1 Scheme of a basic computational model of the MB-inspired network. The input layer is randomly connected with the $\alpha - \beta$ and $\alpha' - \beta'$ lobes that are themselves interconnected by synapses subject to learning. The conditioning layer is finally needed to exploit the information embedded in the lobes, through reward-based learning processes

codifies different intensity of the corresponding feature. Within the same group, neurons are linked together through inhibitory synapses to guarantee that only one neuron in each group remains excited (i.e. winner-takes-all solution). Neurons in different groups are connected using plastic synapses that are reinforced when neurons are firing together, according to the STDP mechanism introduced in Chap. 2.

In the model, each neuron in the AL layer has a probability P = 25% of being connected to the KCs. The choice of the sparse connection between the first and the second layer is directly related to the known topology in the biological counterpart [36]. Probability in flies is an average of 4 out of 150 projection neurons per KC.

5.2.2 Initial Model of the MB Lobes

The KCs in the MBs, as outlined in Chap. 1, project through the peduncle into the lobes. The lobes possess roughly the same topology, but are involved in different functionalities. Our first architecture was restricted to model the structure and functions of the $\alpha - /\beta$−lobes, and the $\alpha' - /\beta'$−lobes, divided into two distinct neural networks. Each network is able to produce a peculiar dynamics: if excited, the neurons in the AL layer begin a competition that leads to the emergence of one cluster of active neurons.

Each lobe was modelled using a lattice of Izhikevich class I neurons with a toroidal topology. The neurons in this layer are all connected to each other according to the paradigm of local excitation and global inhibition.

(a)

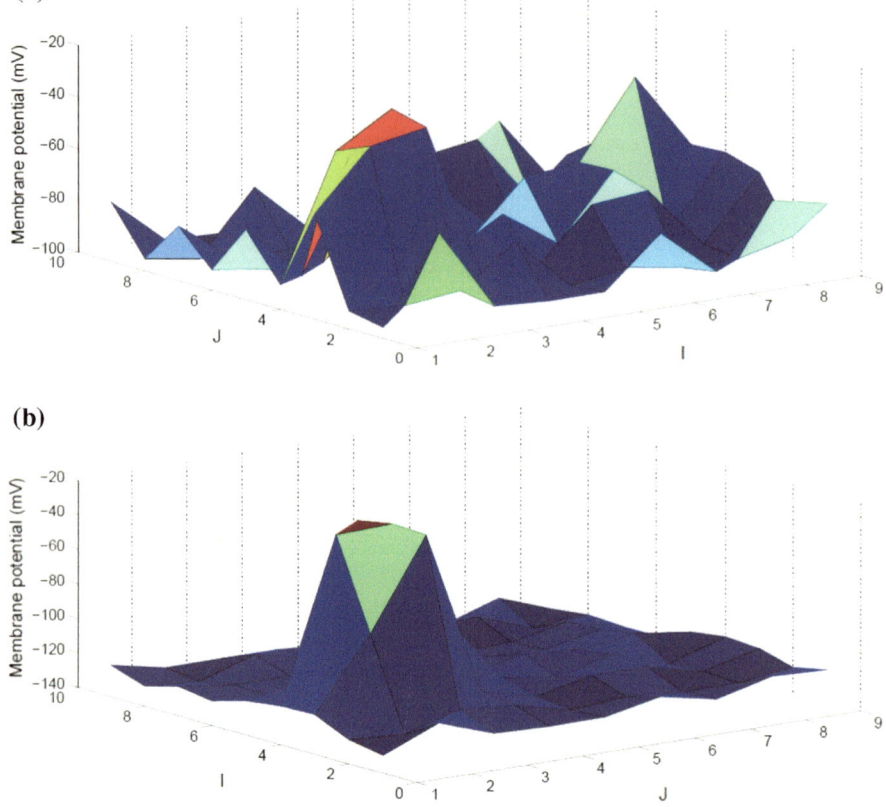

(b)

Fig. 5.2 Cluster formation in the $\alpha - \beta$-lobes. The mean value of the neuron membrane potential is reported for a time window at the beginning of the simulation (**a**) and at the end when a cluster of activity is established (**b**)

The lobes are connected to each other through two sets of synapses, one from the $\alpha - /\beta$−lobes to the $\alpha' - /\beta'$−lobes and vice-versa.

We can assume, on the basis of the biological evidences, that information that reaches the $\alpha' - /\beta' -$ lobes is retained there and stored for memory purposes. In particular we hypothesize that the signals coming from the ALs through the calices are delayed while reaching in the $\alpha' - /\beta'$−lobes.

Under these conditions, the winning cluster in the $\alpha - /\beta$−lobes represents the input presented to the ALs at the actual step, whereas the winning cluster in the $\alpha' - /\beta'$−lobe represents the input presented to the ALs at the previous time step. The synapses between the lobe systems are reinforced when there are two clusters simultaneously active in different lobes. This structure is able to detect whether the object presented as input is the same for two different subsequent acquisitions. In fact, under these conditions, the plastic synapses between the lobes create a positive

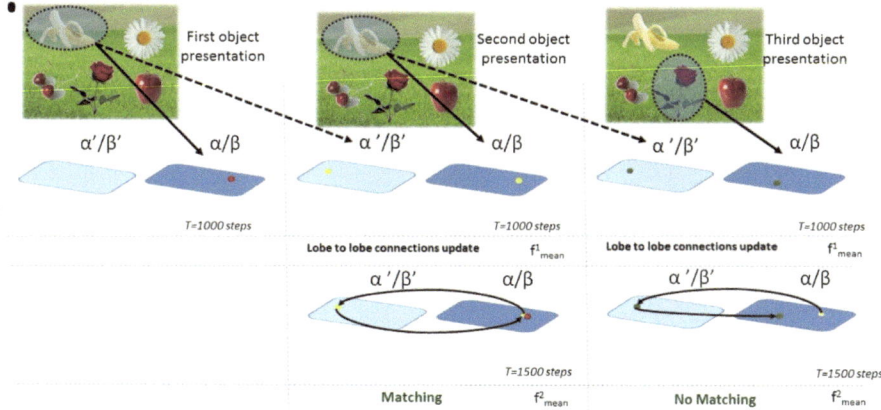

Fig. 5.3 Network evolution due to the presentation of a series of objects. When the first object is presented, its features are processed and a cluster of neural activity arises in the lattice. During the presentation of a second object the $\alpha - /\beta$−lobes behave in a similar way whereas the $\alpha' - /\beta'$−lobes are exited by the lobe-to-lobe connections. If a loop arises, a significant increment in the spiking rate is obtained allowing the matching/no-matching discrimination. This figure was reprinted from [11], © Elsevier 2013, with permission

loop between the clusters in the two lobe systems: as a consequence the spiking rate of the active neurons is increased. We will assume the presence of a neuron sensitive to the firing activity of the $\alpha - /\beta$−lobes network. The sequence of the network evolution is reported in Fig. 5.3. In the first step, two subsequent presentations of the same object generate a positive loop between the two lobe systems that correspond to an increment of spiking rate, whereas during the following presentation, a different object is recognized destroying the loop generated and losing the boosting in the spiking activity within the lobes. In the developed model, the mean spiking activity of the $\alpha - /\beta$−lobes is encoded in a neuron used to discriminate the matching/no-matching events. It is possible to find a threshold in the neural activity of the $\alpha - /\beta$−lobes in order to distinguish the activity in the case of loop and no-loop connection as illustrated in Fig. 5.4.

5.2.3 Premotor Area

Biological evidences discussed in Chap. 1, revealed that context generalisation, visual attention, adaptive termination and decision making are behaviours that involve MBs [27, 44]. Furthermore, MBs have also a role in the control of motor activity. For example, initial motor activity in MB-ablated flies is high, whereas long-term acquisitions show a considerable reduction in motor activity [29].

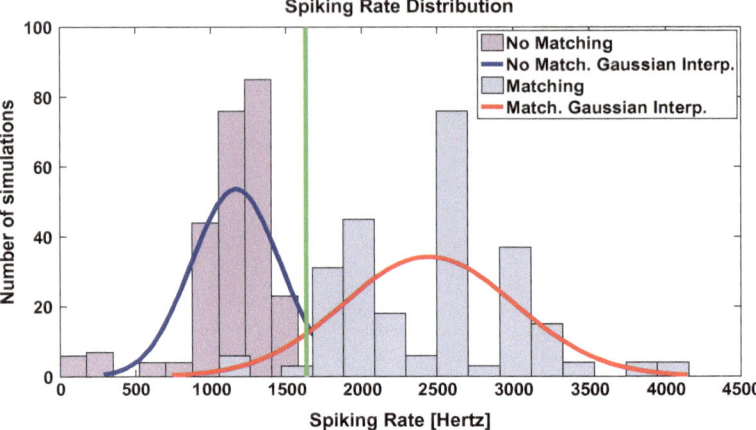

Fig. 5.4 Statistical distribution elaborated using over 500 simulations of the mean spiking activity of the $\alpha - /\beta$—lobes after the reinforcement of the feedback connections between lobes. When a loop is established, the spiking rate is significantly increased and a threshold at $f = 1632$ Hz can be adopted to distinguish the matching/no-matching of two consecutive presented objects with a processing error of about 3%. This figure was reprinted from [11], © Elsevier 2013, with permission

In the developed architecture the activity of the KCs in the MB-lobes is extracted to determine the system behaviour, realizing a connection with the premotor area devoted to the robot control. The MBs and the premotor area are connected via an associative structure that uses the STDP paradigm (see Chap. 2 for details) for a positive/negative-based reinforcement learning.

5.2.4 *Context Layer*

Expectation is the capability of a system to predict the next element on the basis of the last presented one. This one step memory could not be enough to discriminate complex sequences: a memory layer, here called context layer was considered to extend the system capabilities. Experimental results using real data demonstrate that the use of context has three useful consequences: (a) it prevents conflicts during the learning of multiple overlapping sequences, (b) it allows the reconstruction of missing elements in presence of noisy patterns, and (c) it provides a structure for the selective exploration of the learned sequences during the recall phase [14].

A first attempt to develop a Context layer was inspired by the path integration models using the principles of a virtual vectorial sum of the spatial position of the previously emerged clusters creating a spatio-temporal map of contexts. The proposed structure contains a pool of independent neurons spatially distributed in a lattice as illustrated in Fig. 5.5. The horizontal axis indicates the time evolution

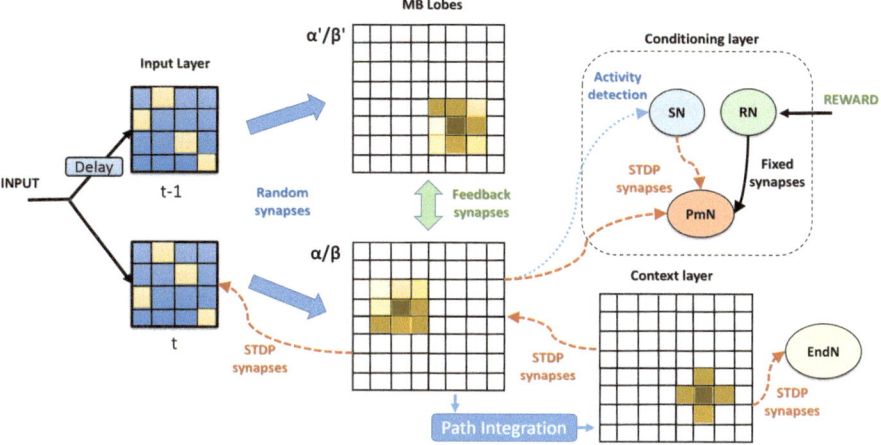

Fig. 5.5 Block scheme of the MB-inspired computational model with the inclusion of the context layer where the history of the sequence is stored using a mechanism similar to path integration. STDP synapses link the active context elements with the next cluster in the $\alpha - \beta$-lobes and the end neuron is used to identify the end of a sequence

whereas the vertical axis represents the internal states sequence, forming the context at each time step. The links between the context layer and the $\alpha - \beta$-lobes are obtained through STDP synapses that realize an all-to-all connections between the two substructures.

Another important element introduced in the architecture is the *End Neuron* (EndN in Fig. 5.5) used to introduce the information about the length of the sequence during the learning phase. Each context neuron can be connected to the end neuron and through STDP can correlate its activation with the end of the sequence. During the testing phase the context activity is reset when the end neuron emits spikes.

The dynamics generated in the context layer resembles the activity produced during a reaction-diffusion process. The idea is to consider the $\alpha - \beta$-lobes randomly connected to the context layer that is composed of groups of neurons topologically organized in lines. This adopted topology resembles the linewise arrangement of the MB fibres and recalls the granule cells in the cerebellum that are responsible for encoding both the pattern of activations and also the time variables involved that will be used by the Kenyon cell to generate the suitable output activity [16].

The process involving the context formation starts when the presented input generates a winning cluster in the $\alpha - \beta$-lobes (time t_0) and consequently the lobes randomly excite the context layer. In this time window, only the first column of context neurons is receptive and a winner-takes-all strategy allows the emergence of a single winning neuron as representative for the current state. After a resetting triggered by the lateral horn, a newly presented element (at time t_1) generates a second cluster in the lobes that randomly excites again the context layer. The previous winner in the context starts a diffusion process with a Gaussian shape toward the second column of neurons. The interactions between these two mechanisms are at the basis of the selection of a second neuron that is related to the history of the previously presented elements. All the neurons in the context are massively connected with the $\alpha - \beta$-lobes by synapses subject to the STDP learning. Therefore the synapses connecting the active neuron that generates the diffusion process, and the current winner in the $\alpha - \beta$-lobes are reinforced. Multiple presentations of the same sequence of elements guarantee that the synapses between the context layer and the $\alpha - \beta$-lobes are strong enough to allow the reconstruction of a learned sequence during the recall phase.

Finally, either rewarding or punishing signals can be linked to the last element of a sequence and this information can determine the selection of the most rewarding sequence to be considered when different possibilities are provided to the system.

5.3   MB-Inspired Architecture: A Step Ahead

The previously described architecture is not able to address additional functionalities recently ascribed to the MBs, like motor learning where the time evolution of reference signals needs to be acquired and reproduced [3]. Therefore the neural structure of some key layers of the previously discussed MB model were improved, developing an architecture able to both classify static features and learn time-dependent signals used as references for the motor system.

The new functionalities taken into consideration need a different processing layer to correctly learn and reproduce spatio-temporal dynamics. We decided to model the KCs' activity as a Liquid State Network (LSN), a lattice of connected spiking neurons similar to a Liquid State Machine [28], that contains mainly local synaptic connections (as in a Cellular Nonlinear Network structure [3, 7] already used for locomotion control [8]), resembling axo-axonal communication among the KCs [17]. The developed lattice elaborates the sensory information, creating a dynamic map, which can be exploited concurrently both for classification and for motor-learning purposes.

From the classification point of view, our architecture exploits the complex internal dynamics that is extracted and condensed in periodic signals whose frequency is able to stimulate specific resonant neurons. We introduced an unsupervised growing mechanism that guarantees the formation of new classes when needed and a supervised learning method to train the read-out maps.

The idea is to consider the context layer as a pool of neurons topologically organized in concentric circles. In bees there is biological evidence that indicates the presence of this kind of arrangement: the calyx neuropil is concentrically organized [43]. Moreover, patterns of genetic expression in DM revealed that KC axons projecting into the γ-lobe form the circumference of the peduncle, whereas a quartet of axon bundles form the core of the peduncle and project into the α- and β-lobes [19]: a concentric axon bundle is part of the fly MBs.

In our model we hypothesize that each ring is stimulated when an input is presented and we assume that the neural activity-wave propagates from the inner to the outer ring in time following a diffusion-like process. The context neurons are connected to the resonant neurons where a competition with the current input information is performed to produce the output of the network through non-elemental learning processes [24, 49] that are an important building block for the expectation and sequence learning processes.

The important role of resonant neurons in classification of auditory stimuli was already discussed in a series of works related to other insect species like crickets [6, 37, 47] and also for the classification of mechanical data provided by a bio-inspired antenna system [35]. The proposed architecture permits to introduce also motor learning capabilities within the MB computational model [3]. Therefore we can both classify static features and learn time-dependent signals to be used to modulate the motor activity.

5.3.1 Network Behaviour for Classification

With regards to classification, each class is represented by a resonant neuron realized through a Morris-Lecar model [31]. A training procedure, using the method introduced in Chap. 2, allows associating a different resonant neuron with each individual input. The read-out map is trained in a supervised way: a periodic wave has been chosen as target signal with a frequency able to stimulate the corresponding resonant neuron. The first read-out map is trained to generate a sine-wave with a frequency of 62.5 Hz when the first input signal is provided. When resonant neurons are not excited by the input, a new resonant neuron with a different frequency is allocated and the corresponding read-out map is learned. The frequency range here adopted spreads from 50 to 250 Hz. This interval was chosen to allow the coexistence of about five different classes. Lower frequencies cannot be considered because at least five periods are used to have a reliable number of spikes in the resonators for a robust classification. More classes can be learned if the time window defined for the target signal is increased accordingly. Moreover, using a wider frequency spectrum, the time constants used in the LSN should be tuned to generate the frequencies needed. A minimum number of spikes (larger than 50% of the maximum allowable depending on the signal frequency) has been considered to determine, whether the resonator is in an active state. The signal coming from the sum neuron is filtered using a Heaviside function, before entering into the resonator. As illustrated in Fig. 5.6, the context layer structure is constituted of concentric rings (only the first three rings were reported for the sake of simplicity). The first ring contains a number of neurons equal to the current number of classes (N_c). In the second ring, for each neuron of the first ring, there are N_c neurons, and the structure develops like a tree in the successive rings. This implies that, within each ring, there are $N_c^{N_r}$ neurons, where N_r is the ring number. The potentially large number of neurons building-up the context layer is justified, because we are simultaneously learning both sequences and

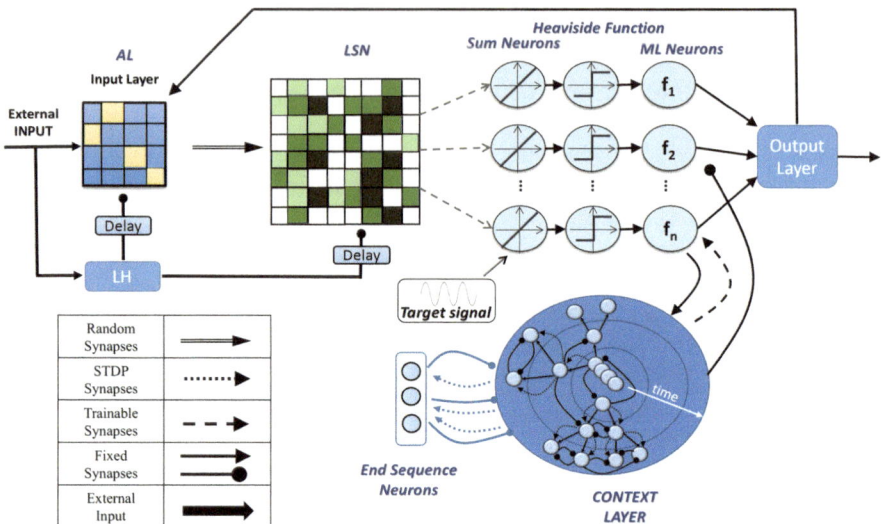

Fig. 5.6 Block scheme of the developed architecture. The external input is processed by the antennal lobes that randomly excite the liquid network; the lateral horn inhibits the lattice activity after a given time window (i.e. 80–100 ms). The liquid activity is modulated with multiple read-out maps that are learned in order to excite specific resonant neurons. The active class neuron (i.e. a resonant Morris-Lecar neuron) stimulates the context layer creating a trace of activities that can successively guide the classification of the next stimulus. Feedback from the context layer to the resonant neurons is subject to learning. A pool of End sequence neurons is also employed to reset the context layer activity. The output layer selects the correct behaviour for the system depending on the winning class and can influence the dynamics of the input layer when conditions like sameness and persistence are identified

sub-sequences. This possibility boosts the capabilities of the structure, much beyond the simple sequence learning. The number of rings present in the context defines the maximum sequence length. Lateral inhibition among neurons of the same ring generates a competition, filtering out potential disturbances [24].

Sequence learning takes place through different stages (i.e. epochs) characterized by a neural activity either stimulated by an external input or an internal input, generated to recall a missing element stored in the network. During each epoch information propagates one ring ahead, from the inner to the outermost.

In every epoch, the winning neuron in the outermost ring and the winner neuron of the previous ring are subject to an STDP learning process which modulates their connection weights as introduced in Chap. 2. In our model, this process could cover larger time scales than the standard STDP. This is required to create correlations among consecutive objects, presentation of which does not happen within the usual STDP time window. Theoretical discussions are presented in [21] whereas further biological evidences are reported in [45] where these learning processes can be modelled using memory traces and reverberation mechanisms [20, 25].

5.3.2 End Sequence Neurons

Each neuron in the context layer is also linked to an end sequence neuron with STDP synapses. In details, all neurons in each ring are connected with the corresponding end sequence neuron arranged in an end sequence vector with a length equal to the number of rings in the context. A rewarding signal, at the end of a sequence, activates the end sequence neuron for the outermost active ring in the context layer. The synapses connecting this end sequence neuron with the winner neuron in the outermost ring will be reinforced accordingly. In our model the reward signal acts as a dopaminergic stimulus on the end sequence neuron to reward the sequence just completed and reset the activity in the context layer for the learning of a new sequence [25].

5.3.3 Neural Models and Learning Mechanisms

Different neuron models were used in the architecture to generate the suitable dynamics needed in the subsystems: the Izhikevich's spiking neurons and the Morris-Lecar model (ML) as introduced in Chap. 2. We adopted the Izhikevich Tonic spiking model in the ALs, context layer and end sequence neurons whereas the Class I model was exploited in the LSN.

A decay rate has been introduced to consider dynamically changing environments where the learned sequences could be forgotten if no longer rewarding. Details on applications of this learning paradigm to biorobotics are illustrated in [9]. Spiking neurons in the KC lattice are fully connected to the sum neurons via plastic trainable synapses. A simple supervised learning method based on the pseudo-inverse algorithm [26] has been adopted and compared with an incremental learning rule as illustrated in [30] where the spiking activity of the neurons is transformed in continuous signals using different functions allowing the evaluation of an error needed for the learning. Different supervised learning methods, based on back propagation for spiking networks [22], have not been adopted due to the presence of recurrent connections in the lattice. Although different learning approaches could be taken into consideration [15], we introduced a simple incremental learning strategy based on the least mean square algorithm, that adapts the synaptic weight using the computed error and the local activity generated by the pre-synaptic neuron, working with the synaptic response (i.e. continuous variables) instead of with the spike train.

The exit condition for the learning process is obtained monitoring the spikes emitted by the ML neurons: when a given number of spikes is correctly emitted, the learning is stopped and the weights of the read-out map are stored in the architecture. To evaluate the performance of the learning process, a comparison between the pseudo-inverse (standard solution) and the biologically more plausible step-by-step method was performed for a simple interpolation task. The results, shown in Fig. 5.7, illustrate that using an incremental procedure, the learning process converges on an oscillatory signal able to properly stimulate the corresponding ML neuron; in fact, the maximum amount of allowable spikes is emitted. It can be noticed that even if

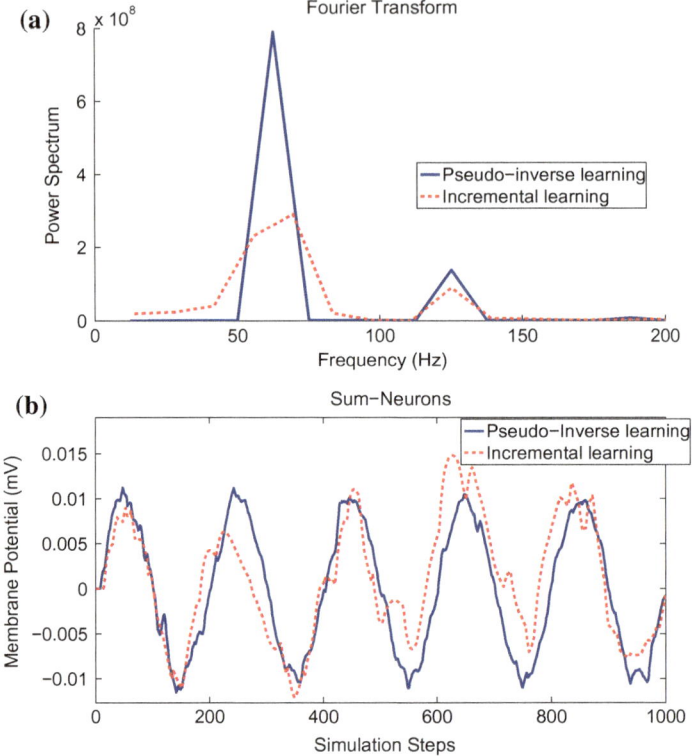

Fig. 5.7 Comparison between the results of the read-out map obtained using a pseudo-inverse (solid line) and an incremental (dashed line) learning method. The power spectrum of the signals (**a**) and the cumulative output of the sum neuron (**b**) are reported

the output signal of the sum neuron does not exactly match the sinusoidal target, the classification phase is successfully accomplished.

5.3.4 Decision-Making Process

The architecture is able to store and consequently retrieve multiple sequences that can be followed in a decision-making task. A rewarding signal, provided to the system at the end of each sequence, is used to evaluate the importance of each stored sequence. In our model the level of reward associated with the generated sequence is directly related to the number of spikes emitted by the end neuron. As an example after learning two sequences *ABBA* and *BDDC*, the system needs to choose which sequence to generate in front of the presentation of the first objects of the two sequences (*A* and *B*). This is a typical case of landmark sequence following to reach a target place. The system internally simulates one and then the other sequence and compares the number of spikes of the end sequence neuron in both cases. In the reported example the sequence *BDDC* is the most rewarding one and then the behaviours related to this sequence are performed.

5.3.5 Learning Sequences and Sub-sequences

The developed architecture, thanks to the Context layer topology, allows to learn not only a sequence of elements but also the sub-sequences included therein. We can exploit this capability offered by the model, allowing each presented element to activate, besides a neuron included in the ring after the last active one, also a neuron in the first ring of the Context.

This procedure allows to consider each element of a sequence as a starting point for a new one. This leads to generation and storage of multiple chains of context activity: under presence of very noisy and complex conditions this strategy could be useful to retain only statistically relevant sub-sequences that are reinforced depending on the presentation frequency that is used to improve the sequence reliability.

An example of learning for sub-sequences is shown in Fig. 5.8, where the evolution of the context layer is shown, while learning the sequence *ABCD*. Each element of

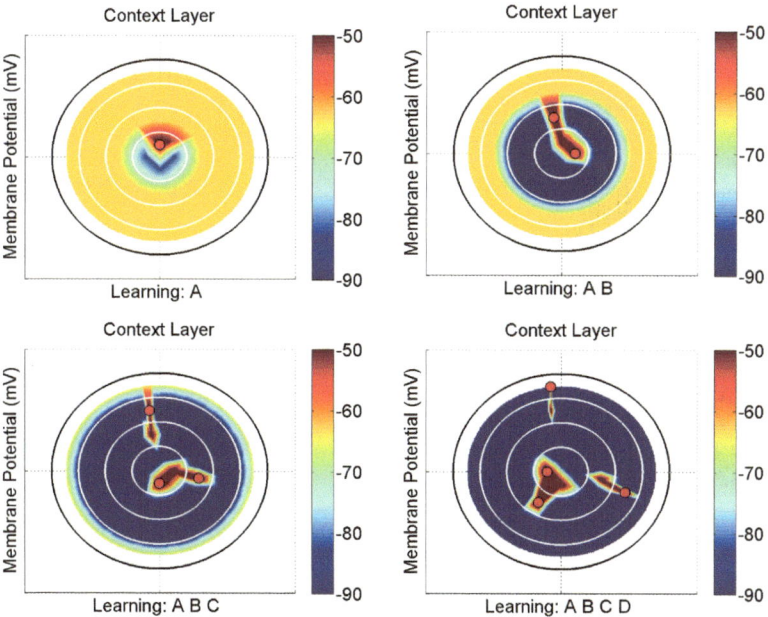

Fig. 5.8 Activity of the Context layer when the sequence *ABC* is provided and the sub-sequences are learned by the system. The red circles represent the active neurons, multiple chains are generated to trace the subsequences *BCD*, *CD* and *D* acquired in the Context layer

the sequence creates a trace in the Context layer that starts from the inner ring and propagates towards the outer ones. The memory trace in the Context layer is incrementally strengthened through multiple presentations of the same sequence.

5.4 Robotic Experiments

To evaluate the performance of the designed architecture, a series of experiments were carried out with a roving platform. The robot is equipped with a PC on board that communicates with a series of micro-controller-based boards used for the motor control. The sensory system consists of two ultrasound sensors used to detect frontal

Fig. 5.9 a Persistence experiment with a roving robot. Trajectory followed by the robot during the experiment. The images acquired from an on-board fish-eye camera are reported for three different processing steps. The robot is able to persistently move in the direction of the inverted T also when a distracter is presented (i.e. the circle). After multiple presentations, the memory associated with the inverted T fades out and the robot follows the circle. **b** Distraction experiment with a rover. The presentation of a new object at step 12 is enough to change the robot behaviour independently of the time/energy already invested in following the previous target. This figure was reprinted from [5], © Elsevier 2015, with permission

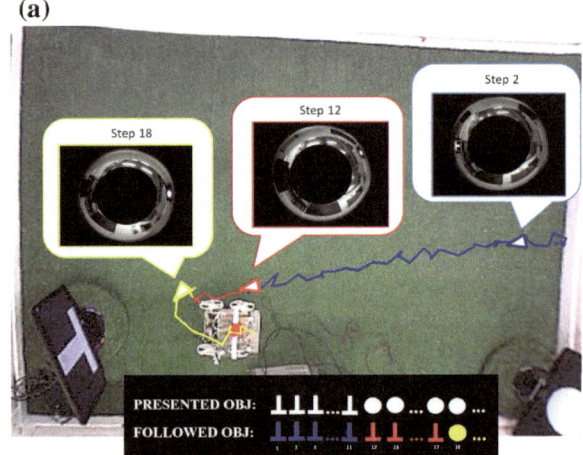

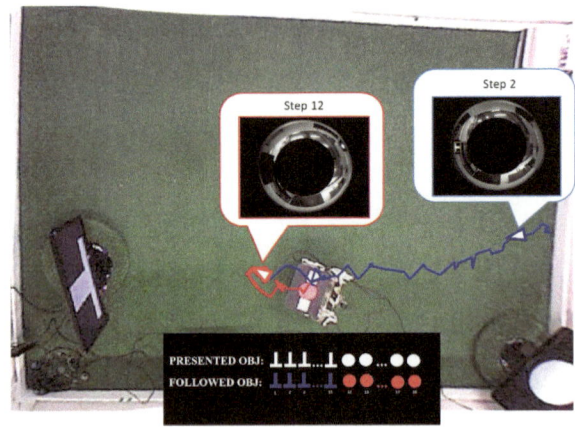

obstacles and an omni-directional camera included to identify the presence of visual targets with specific shapes in the environment (i.e. objects of interest). A first experiment is directly related to the persistence behaviour as shown in Fig. 5.9a. The robot is attracted by the objects shown on the monitors placed in the arena and is able either to filter out the distracter, replicating the attention capability present in the wild-type insect, or to switch among the presented objects, like in the MB-defective case, as reported in Fig. 5.9b, where the robot is continuously distracted by the presentation of the conflictual stimuli.

Honeybees are able to deal with a maze by using symbolic cues [18, 50], and ants are able to navigate following routes [13], therefore we evaluate our control architecture on a roving robot facing with scenarios where the available paths from the entrance to the exit of a maze, have to be acquired and evaluated [51]. The objective is to identify from the visual input the landmarks of interest and process the acquired landmark features using a spiking neural network to learn multiple sequences of events/actions and the corresponding expected rewards.

The information acquired through the visual system is pre-processed using simple segmentation libraries, and then processed by the network to identify the suitable

Fig. 5.10 Robotic experiments performed using different maze configurations for the learning (**a**) and the testing (**b**) phase. Both the top and lateral view acquired from the starting point are reported. The solid line represents the trajectory followed by the robot. The scenario is equivalent to the biological experiments performed with honeybees [51]. During the learning phase the robot memorizes the correct actions to be associated with each presented element. During the testing phase (**b**) the robot can solve a different maze using the landmark projected on a single monitor

action (or more complex behaviours) to be performed. In the proposed experiment, the robot can perform a turning action followed by a forward motion to reach a new branch of the maze. Two monitors are used to simulate the presence of landmarks in each branch of the multiple T-maze. To guide the robot in the maze we considered a double reversed T to indicate a right turn and a double circle for a left turn. During the learning phase, the correct actions to be performed are provided showing only one landmark each time on the monitor positioned in the correct turning direction (see Fig. 5.10a). During the testing phase only one monitor is used and the structure of the maze is modified to demonstrate that the robot is able to solve it using the knowledge acquired in the previous phase (Fig. 5.10b).

These results show that the robot is able to negotiate a maze by using symbolic cues as shown by honeybees [51]. We then included the sequence learning capabilities considering a more complex scenario. The robot initially learns two sequences of actions to be performed to solve the maze following two different routes that guarantee, at the end, reward signals with different values. The first sequence (i.e. $Inverted - T$, $Circle$, $Inverted - T$) is associated to the left, right and then left maze-branch selection, whereas the second learned sequence (i.e. T, $inverted - T$) is associated to a right and then left turning action. The reward level that modulates the stimulation of the end sequence neuron, is lower for the first sequence than for the second one that should be preferred. During the testing phase the robot is placed in front of two concurrent stimuli (i.e. T and $inverted - T$) to perform a decision-making process. In this way the architecture can internally retrieve differ-

Fig. 5.11 After learning two sequences, the robot selects the most rewarding one performing a decision-making process. The learned sequences are depicted on the two sides; the solid line represents the selected trajectory during the testing phase

ent sequences, depending on the initial stimulus, and selects the most performing one depending on the spiking activity of the end neuron, that encodes the cumulative level of rewards obtained, for that sequence, during the learning process. The control structure internally simulates the outcome of the two possible sequences to be followed and, on the basis of the expected reward, selects the most rewarding one (Fig. 5.11).

5.5 Conclusions

The ability to understand the environment is a fundamental skill for living beings and needs to be acquired as a dynamic process. The context in which events occur cannot be ignored, in fact, sometimes it is more important than the events themselves. Starting from the biological evidences concerning insect capabilities to learn sequences of events and the known facts on the neural structures responsible for these processes, in this work a neural-based architecture for sequence representation and learning is proposed. The proposed model is therefore based on experimental evidences on insect neurodynamics and on specific hypotheses on the mechanisms involved in the processing of time-related events. Starting from basic capabilities like attention, expectation and others, the MB-inspired model was extended to include sequence learning with the addition of a context layer. A series of implementations is proposed: simulation and experimental results using a roving robot, demonstrate the effectiveness of the proposed architecture that represents a key structure for the development of a complete insect brain computational model. In conclusion, the model architecture discussed here represents a fundamental building block toward an artificial neural processing structure unifying different functionalities, and performing different behaviours, that biological counterparts are able to show.

References

1. Adeli, H., Park, H.: A neural dynamics model for structural optimization—theory. Comput. Struct. **57**(3), 383–390 (1995)
2. Ahmadkhanlou, F., Adeli, H.: Optimum cost design of reinforced concrete slabs using neural dynamics model. Eng. Appl. Artif. Intell. **18**(1), 65–72 (2005)
3. Arena, P., Caccamo, S., Patanè, L., Strauss, R.: A computational model for motor learning in insects. In: IJCNN, pp. 1349–1356. Dallas, TX (2013)
4. Arena, P., Calí, M., Patanè, L., Portera, A.: A fly-inspired mushroom bodies model for sensory-motor control through sequence and subsequence learning. Int. J. Neural Syst. **24**(5), 1–16 (2016)
5. Arena, P., Calí, M., Patanè, L., Portera, A., Strauss, R.: Modeling the insect mushroom bodies: application to sequence learning. Neural Netw. **67**, 37–53 (2015)
6. Arena, P., Fortuna, L., Frasca, M., Ganci, G., Patanè, L.: A bio-inspired auditory perception model for amplitude-frequency clustering. Proc. SPIE **5839**, 359–368 (2005)
7. Arena, P., Fortuna, L., Frasca, M., Patanè, L.: Sensory feedback in CNN-based central pattern generators. Int. J. Neural Syst. **13**(6), 349–362 (2003)

8. Arena, P., Fortuna, L., Frasca, M., Patanè, L.: A CNN-based chip for robot locomotion control. IEEE Trans. Circ. Syst. I **52**(9), 1862–1871 (2005)
9. Arena, P., Fortuna, L., Frasca, M., Patanè, L.: Learning anticipation via spiking networks: application to navigation control. IEEE Trans. Neural Netw. **20**(2), 202–216 (2009)
10. Arena, P., Patanè, L., Termini, P.: Learning expectation in insects: a recurrent spiking neural model for spatio-temporal representation. Neural Netw. **32**, 35–45 (2012)
11. Arena, P., Stornanti, V., Termini, P., Zaepf, B., Strauss, R.: Modeling the insect mushroom bodies: Application to a delayed match-to-sample task. Neural Netw. **41**, 202–211 (2013)
12. Aso, Y., Hattori, D., Yu, Y., Johnston, R.M., Iyer, N.A., Ngo, T.T., Dionne, H., Abbott, L., Axel, R., Tanimoto, H., Rubin, G.M.: The neuronal architecture of the mushroom body provides a logic for associative learning. eLife **3** (2014). https://doi.org/10.7554/eLife.04577
13. Baddeley, B., Graham, P., Husbands, P., Philippides, A.: A model of ant route navigation driven by scene familiarity. PLoS Comput. Biol. **8**(8), e1002,336 (2012). https://doi.org/10.1371/journal.pcbi.1002336
14. Berthouze, L., Tijsseling, A.: A neural model for context dependent sequence learning. Neural Process. Lett. **23**(1), 27–45 (2006)
15. Brea, J., Senn, W., Pfister, J.: Matching recall and storage in sequence learning with spiking neural networks. J. Neurosci. **33**(23), 9565–9575 (2013)
16. Buonomano, D.V., Mauk, M.D.: Neural network model of the cerebellum: temporal discrimination and the timing of motor responses. Neural Comput. **6**(1), 38–55 (1994). https://doi.org/10.1162/neco.1994.6.1.38
17. Cassenaer, S., Laurent, G.: Hebbian STDP in mushroom bodies facilitates the synchronous flow of olfactory information in locusts. Nature **448**(7154), 709–713 (2007)
18. Collet, T., Fry, S., Wehner, R.: Sequence learning by honeybees. J. Comp. Physiol. A. **172**(6), 693–706 (1993)
19. Davis, R., Han, K.: Neuroanatomy: mushrooming mushroom bodies. Curr. Biol. **6**, 146–148 (1996)
20. Drew, P.J., Abbott, L.F.: Extending the effects of spike-timing-dependent plasticity to behavioral timescales. Proc. Nat. Acad. Sci. USA **103**(23), 8876–8881 (2006)
21. Friedrich, J., Urbanczik, R., Senn, W.: Code-specific learning rules improve action selection by populations of spiking neurons. Int. J. Neural Syst. **24**(05), 1450,002 (2014). http://www.worldscientific.com/doi/abs/10.1142/S0129065714500026
22. Ghosh-Dastidar, S., Adeli, H.: Improved spiking neural networks for eeg classification and epilepsy and seizure detection. Int. J. Neural Syst. **14**(03), 187–212 (2007)
23. Ghosh-Dastidar, S., Adeli, H.: Spiking neural networks. Int. J. Neural Syst. **19**(4), 295–308 (2009)
24. Giurfa, M.: Cognitive neuroethology: dissecting non-elemental learning in a honeybee brain. Curr. Opin. Neurobiol. **13**(6), 726–735 (2003)
25. Izhikevich, E.M.: Solving the distal reward problem through linkage of stdp and dopamine signaling. Cereb. Cortex **17**, 2443–2452 (2007)
26. Jaeger, H.: Short term memory in echo state networks. GMD-Report German National Research Institute for Computer Science **152** (2002)
27. Liu, L., Wolf, R., Ernst, R., Heisenberg, M.: Context generalization in Drosophila visual learning requires the mushroom bodies. Nature **400**, 753–756 (1999)
28. Maass, W., Natschlaeger, T., Markram, H.: Real-time computing without stable states: a new framework for neural computation based on perturbations. Neural Comput. **14**(11), 2531–2560 (2002)
29. Martin, J.R., Ernst, R., Heisenberg, M.: Mushroom bodies suppress locomotor activity in Drosophila melanogaster. Learn. Memory **5**, 179–191 (1998)
30. Mohemmed, A., Schliebs, S., Matsuda, S., Kasabov, N.: Span: Spike pattern association neuron for learning spatio-temporal spike patterns. Int. J. Neural Syst. **22**(04), 1250,012 (2012). http://www.worldscientific.com/doi/abs/10.1142/S0129065712500128
31. Morris, C., Lecar, H.: Voltage oscillations in the barnacle giant muscle fiber. Biophys. J. **35**, 193–213 (1981)

32. Mosqueiro, T.S., Huerta, R.: Computational models to understand decision making and pattern recognition in the insect brain. Curr. Opin. Insect Sci. **6**, 80–85 (2014)
33. Nowotny, T., Huerta, R., Abarbanel, H., Rabinovich, M.: Self-organization in the olfactory system: one shot odor recognition in insects. J. Comput. Neurosci. **93**, 436–446 (2005)
34. Nowotny, T., Rabinovich, M., Huerta, R., Abarbanel, H.: Decoding temporal information through slow lateral excitation in the olfactory system of insects. J. Comput. Neurosci. **15**, 271–281 (2003)
35. Patanè, L., Hellbach, S., Krause, A.F., Arena, P., Duerr, V.: An insect-inspired bionic sensor for tactile localisation and material classification with state-dependent modulation. Frontiers Neurorobotics **6**(8) (2012). http://www.frontiersin.org/neurorobotics/10.3389/fnbot.2012.00008/abstract
36. Perez-Orive, J., Mazor, O., Turner, G., Cassenaer, S., Wilson, R., Laurent, G.: Oscillations and sparsening of odor representations in the mushroom body. Science **297**, 359–365 (2002)
37. Russo, P., Webb, B., Reeve, R., Arena, P., Patanè, L.: Cricket-inspired neural network for feedforward compensation and multisensory integration. In: IEEE Conference on Decision and Control (2005)
38. Sachse, S., Galizia, C.: Role of inhibition for temporal and spatial odor representation in olfactory output neurons: a calcium imaging study. J. Neurophysiol. **87**, 1106–1117 (2002)
39. Scherer, S., Stocker, R., Gerber, B.: Olfactory learning in individually assayed Drosophila larvae. Learn. Memory **10**, 217–225 (2003)
40. Schmuker, M., Pfeil, T., Nawrot, M.P.: A neuromorphic network for generic multivariate data classification. Proc. Nat. Acad. Sci. USA **111**(6), 2081–2086 (2014). https://doi.org/10.1073/pnas.1303053111. http://www.pnas.org/content/111/6/2081.abstract
41. Smith, D., Wessnitzer, J., Webb, B.: A model of associative learning in the mushroom body. Biol. Cybern. **99**, 89–103 (2008)
42. Stocker, R., Lienhard, C., Borst, A.: Neuronal architecture of the antennal lobe in *Drosophila melanogaster*. Cell Tissue Res. 9–34 (1990)
43. Strausfeld, N.J.: Organization of the honey bee mushroom body: Representation of the calyx within the vertical and gamma lobes. J. Comparative Neurology **450**(1), 4–33 (2002). 10.1002/cne.10285. http://dx.doi.org/10.1002/cne.10285
44. Tang, S., Guo, A.: Choice behavior of Drosophila facing contradictory visual cues. Science **294**, 1543–1547 (2001)
45. Tanimoto, H., Heisenberg, M., Gerber, B.: Experimental psychology: event timing turns punishment to reward. Nature **430**, 983 (2004)
46. Turner, G., Bazhenov, M., Laurent, G.: Olfactory representations by *Drosophila* mushroom body neurons. J. Neurophysiology, 734–746 (2008)
47. Webb, B., Wessnitzer, J., Bush, S., Schul, J., Buchli, J., Ijspeert, A.: Resonant neurons and bushcricket behaviour. J. Comparative Physiol. A **193**(2), 285–288 (2007). https://doi.org/10.1007/s00359-006-0199-1
48. Wehr, M., Laurent, G.: Odor encoding by temporal sequences of firing in oscillating neural assemblies. Nature **384**, 162–166 (1996)
49. Wessnitzer, J., Young, J., Armstrong, J., Webb, B.: A model of non-elemental olfactory learning in Drosophila. J. Neurophysiol. **32**, 197–212 (2012)
50. Zhang, S., Bartsch, K., Srinivasan, M.: Maze learning by honeybees. Neurobiol. Learn. Mem. **66**(3), 267–282 (1996)
51. Zhang, S., Si, A., Pahl, M.: Visually guided decision making in foraging honeybees. Frontiers Neurosci. **6**(88), 1–17 (2012)

Chapter 6
Towards Neural Reusable Neuro-inspired Systems

6.1 Introduction

In order to design and realise bio-inspired cognitive artefacts, a deep understanding of the mammalian brain would be necessary, which is still lacking. On the other hand, even the most sophisticated robot does not possess the ambulatory or cognitive capabilities of insects like ants, bees or flies. Already here an open question on the actual working of neurons and neuron assemblies arises: how can such tiny insect brains cope in such an efficient way with the high number of parallel exogenous and multimodal stimuli they are exposed to? A straightforward answer would be to try, exploiting results of focussed experiments, to build a block-size model made-up of interconnected sub-systems. Each block in this model would represent a specific part of the insect brain related to a specific behaviour and one would try to build a meaningful network of connections among them. The final figure that arises from this strategy is to build a cooperative-competitive dynamical system which would combine all the modelled behaviours, designed and implemented basically as a super-position of dissociable functional components. This is indeed what is already done in literature and the approach is powerful and valid for modelling most of the sensory-motor direct pathways and related processing. Some recent European projects were focussed toward building an insect brain computational model based on such type of approach [2]. But it is true that complex behaviours need a number of tasks to be implemented concurrently, not in competition, but rather in cooperation. Moreover, studying in details the peculiarities of the neural structures in charge of eliciting particular behaviours, an idea was conceived that a block-size perspective in modelling the insect brain could be considered somewhat restrictive, since it potentially could not exploit the richness of dynamics generated in particular substructures of the brain. Particularly the dynamics of the MBs should be efficiently exploited in other parts of the brain to generate complex proto-cognitive behaviours. MBs, often homologized with the mammalian limbic system, where ancestral perception and emotions are forged, are the centre where learning mechanisms were first identified

© The Author(s) 2018
L. Patanè et al., *Nonlinear Circuits and Systems for Neuro-inspired Robot Control*,
SpringerBriefs in Nonlinear Circuits, https://doi.org/10.1007/978-3-319-73347-0_6

in the insect brain. There, memories are formed and used to build sensory-motor associations. From these facts the idea emerged to look at the insect brain from a holistic view. The complex dynamics present in some parts of the insect brain, like the MBs, could be seen as a place where complex spatio-temporal patterns of neural activity are formed, to be used concurrently for a number of different behavioural responses. This will happen presumably in cooperation with other parts of the brain which can exploit such dynamics and can, in principle, be much simpler than the MBs. This idea has a lot of features in common with the concept of *neural reuse*, recently introduced by Anderson [1]: central neural lattices could be massively re-used for contributing to different behaviours, thanks to variable connections. This concept, drawn from a high level, psychological perspective, can be reflected into a low level view: the same neural structure can have different uses according to which part of its dynamics is exploited to match the outer-world needs. For instance, locomotion control in a legged robot is a particular case of ground manipulation and implies concurrent motion of different parts of the body. On the other side, perception-for-action implies an abstract manipulation of the surrounding environment to reach an ultimate goal. Examples of this concept can be found in the insect world: cockroaches can easily shift from basic hexapedal to a fast running bipedal locomotion, whereas dung beetles perform manipulation and transportation with front and mid legs while walking with the remaining two legs. To conceive new paradigms for the representation of these behaviours from a neural circuit viewpoint, we have to refer to known facts from the insect brain processing areas. Insect data are primarily available from physiology (including also microsurgery) and behaviour research in honeybees, grasshoppers and cockroaches. For *Drosophila* flies rich data based on various techniques like calcium imaging, neurogenetic manipulation, or biochemistry of plasticity (Sect. 1.3), are also available. MBs are the necessary centres for olfactory classical conditioning and are the site for keeping aversive and appetitive memories of odours on different time scales. They have later been implicated in other forms of learning (visual, motor, classical and operant conditioning) and classification tasks (visual foreground-background discrimination, disambiguation of contradictory visual cues), also in motivation control [14] and sleep regulation. Very recent methods in *Drosophila* neurogenetics [4] allow to genetically drive the activity, or silence the output, of single cell types for most, if not all of the MB neuron types individually.

Taking inspiration from *Drosophila melanogaster*, and from insects in general, their most important exteroceptive senses, used to acquire information from the environment, are olfactory and visual perception. Olfactory and also visual processing are two fundamental sensory modalities used by insects for food retrieval and predator escaping. The fruit fly detects food sources and identifies dangerous substances mainly through its olfactory system. The sensory information acquisition starts from the antenna where several different chemical receptor types are located. Subsequently through the antennal lobe the acquired signals are conveyed via the projection neurons to the mushroom bodies (MBs) where the mechanisms of learning and memory for odour classification and behaviour association have been identified.

Besides the peripheral processing centres for vision and olfaction, the MBs are the insect central brain structure with the best-described wiring and physiology; they are key structures for different forms of insect learning. Judged by the expression pattern of early neurodevelopmental genes, MBs are the insect orthologue of the human hippocampus (underlying declarative learning and spatial memory) and amygdala (underlying fear, emotions and the memory for hazards) [17].

The other sensory modality, massively used by the fruit fly to orient itself in the environment, is vision.

Visual stimuli are acquired by the compound eyes and, after processing by the optic lobes, they reach, via the optic tubercles, the central complex. Here particularly the protocerebral bridge, the fan-shaped body and the ellipsoid body are the substructures involved in visual orientation [18] and in visual learning mechanisms based on classical conditioning [12].

Notably MBs and CX are the most important and well-studied insect brain centres and only recently direct connections among them were identified (see Chap. 1). Another important sensory modality less studied in *Drosophila* but deeply investigated in other insects like stick insect [6] and cockroach [19] is related to the tactile sense. Mechanosensors are distributed all over an insect's body and allow a direct interaction with the environment during walking and also during flying. Of particular importance are mechanosensors in the body appendages, foremost in the antennae. Due to the massive presence of mechanoreceptors, the processing of this information is performed in a decentralized way and to large parts in the thoracic ganglia (part of the central nervous system) to create local feedback for reflexes and other simple reactive strategies. Nevertheless as for the other sensory modalities, part of the acquired information is transferred to the central brain and very recently, experiments show that the CX is involved in handling this information. In particular the PB seems to be responsible for orientation strategies in presence of tactile stimuli. These information are the basic facts that show how different neural centres cooperate in performing different behavioural tasks controlling the same neural circuits. Cooperation and exploitation of multiple complex dynamics is then the common denominator of this new way of looking at the insect brain.

6.2 Multimodal Sensory Integration

The different sensory modalities present in insects are processed following distinct pathways from the periphery to the central brain even if some local shortcuts can be observed in the case of reflexes. The interaction among the modalities is needed to have the complete portrait used to perform behavioural choices.

A scheme of the most important processing stages for each sensory modality in *Drosophila* is shown in Fig. 6.1. The sensory-motor loop mainly involves the MBs for olfactory stimuli whereas the CX is strongly involved in the processing of visual and tactile stimuli. For all the reported paths, a central and a peripheral processing phase can be distinguished.

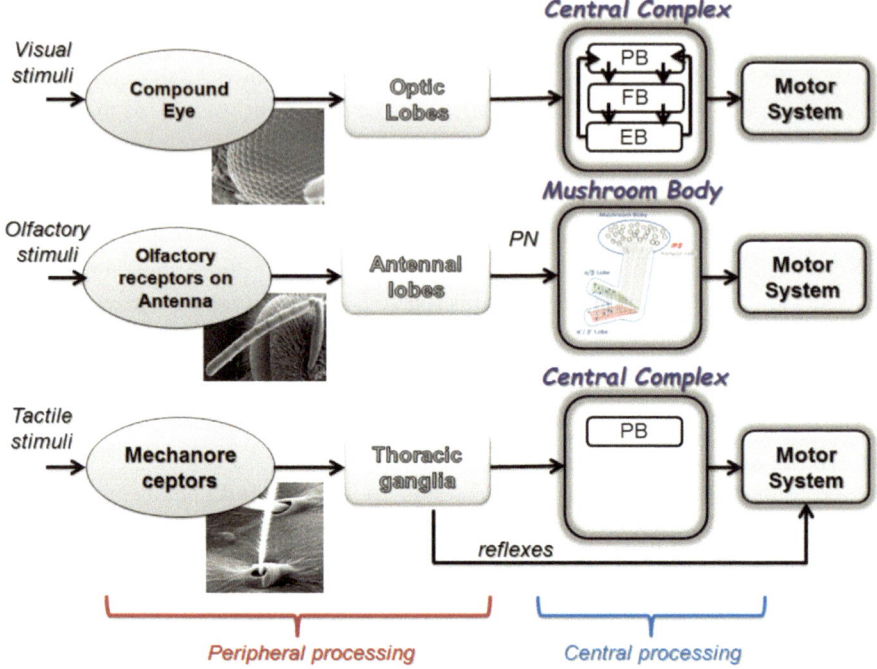

Fig. 6.1 Block scheme of the centres involved in the management of different sensory modalities: visual, olfactory and tactile. A central and a peripheral portion of processing is present in the identified sensory paths

A detailed scheme of the neural centres involved in the processing with particular attention on the different orientation behaviours triggered by the available sensory modalities or combinations thereof is depicted in Fig. 6.2. The different sensory receptors are distributed along the body and the information acquired is processed by several interconnected neural centres. Considering the compound eye, the photo-receptors acquire visual information that is transferred via the optic lobes to the CX that can trigger motor activities by sending commands to pre-motor areas. The antennae include multiple receptors processed by distinct centres. Chemical receptors located in the antennae are connected to the antennal lobes that, via the projection neurons transfer the acquired signals to the lateral horns (LH) and to the MB calyces. The LH is directly connected to the pre-motor area for inborn responses to olfactory stimuli [15] and provides a delayed event-triggered resetting signal to the lobe system [13]. On the other hand the MB lobe system provides feedback to the AL to filter the input creating mechanisms like attention and expectation. Conditioned learning can be obtained both in MB and CX using unconditioned stimuli from octopaminergic and dopaminergic neurons. The antennae are also the place of mechanoreceptors as for instance the Johnston's organs (JO), the fly's antennal hearing apparatus, used for sound and wind detection. The neural centre involved in the processing is

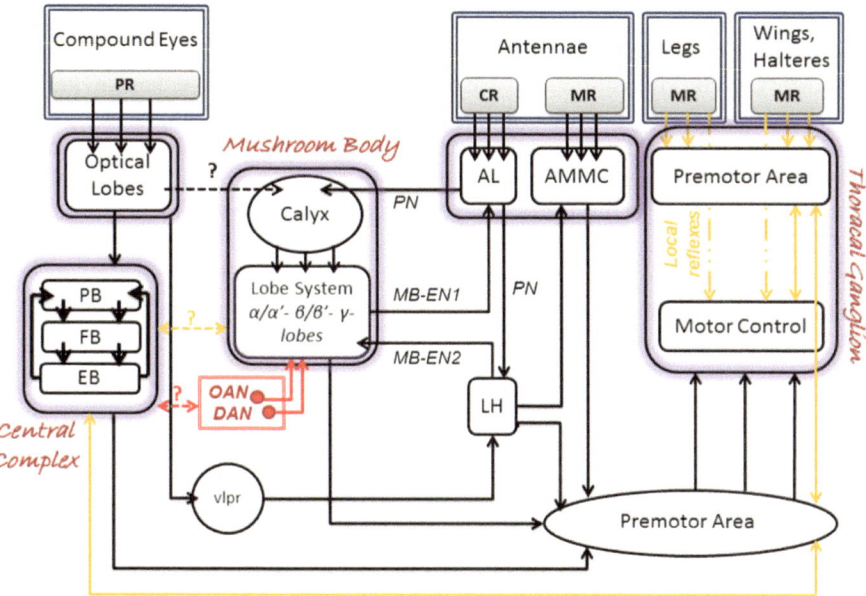

Fig. 6.2 Block-size model of the neural centres involved in orientation behaviour triggered by different sensory modalities: visual, olfactory and tactile. Visual stimuli are acquired by the compound eyes via photo-receptors (PR) and, after a pre-processing at the level of the optic lobes, are handled by the CX that decides the motor commands to allow a visual orientation behaviour. Olfactory stimuli are acquired by the antennae via chemical-receptors (CR) and through the antennal lobes (AL) and the projection neurons (PN) reach the MBs where mechanisms of learning and memory are performed through reinforcement learning signals via dopaminergic neurons (DAN) and indirectly via octopaminergic ones (OAN). Tactile stimuli are acquired by mechanoreceptors (MR) distributed along the whole body. High numbers are found on antennae, legs and the halters which are used for flight control. The information coming from sensors distributed on legs and on the main body, can be processed locally by the thoracic ganglia for the generation of local reflexes, however, an interaction with the CX is also envisaged and in particular the PB seems to be involved in orientation strategies based on mechano-sensory modality

the antennal mechanosensory and motor centre (AMMC). The interaction between visual and olfactory-based orientation is obtained via the ventrolateral protocerebrum (vlpr) that mediates the transfer of parts of the visual information to the LH. This multimodal sensory integration is confirmed also by experiments where, in absence of a well contrasted visual panorama, olfactory orientation is not obtained [9]. Finally, mechanoreceptors placed in the main body and on the legs are handled by the thoracic ganglia where local reflexes can be performed. Moreover, part of this information can be transferred to the CX and in particular the role of the protocerebral bridge seems to be fundamental for mechanosensory-based orientation.

Duistermars and Frye analyzed the interaction between wind-induced and odour-induced orientation [9]. The presence of wind passively rotates the aristae on the antennae, stimulating the JO neurons that, via the AMMC trigger an upwind orientation behaviour. In presence of an odour gradient, the olfactory receptor neurons on the antennae trigger an orientation response from the MB/AL system and the LH-AMMC connection allows the use of the same orientation mechanism know for wind-induced behaviour. In fact the AMMC, via the antennal motor neurons triggers the movement of the aristae that activates the previously introduced orientation mechanism. This example shows, how apparently different orientation strategies can use the same neural circuitries for the generation of the behavioural response.

The known information about the sensory processing phase in the fly will be discussed in the following with particular attention to mechano-sensors that will have a significant impact on the implementation of locomotion control systems.

6.3 Orientation: An Example of Neural Reuse in the Fly Brain

Orientation is a fundamental requirement for insects that need to use sensory information to reach food sources or to avoid dangerous situations. Different sensory modalities can use the same pathways for eliciting motor actions, so as to avoid the duplication of neural mechanisms that can be easily shared. The main facts known from biology are reported below.

6.3.1 Role of CX in Mechanosensory-Mediated Orientation

To understand the role of higher brain centres in handling mechanosensory information, different experimental campaigns were performed. The idea was to limit the variables responsible for the fly behaviour only to mechanosensory information. A scenario taken into consideration consists of a fly, forced to move in the forward direction when an air shot is used to stimulate the movement. The wild-type fly behaviour consists of a change of its direction of motion whereas mutant flies of the CX are less reactive to the shot and quite continuous in performing the on-going behaviour. Going deeper into details through the involvement of several mutant lines, the central role of the PB emerged. Experiments with PB mutant flies show, how an even tiny defect in part of the PB drastically modifies the fly's orientation behaviour in response to these tactile stimuli.

It is interesting to notice that the fly's CX is a well-known centre for orientation in insects, deeply studied with respect to the visual sensory modality. In practice, there is a topological neural mapping between the fly's visual field and the particular PB glomerular structure. On the other side, in new experiments an intact PB is needed to perform a suitable orientation behaviour in response to mechanical stimulation.

This fact is of fundamental importance in view of the concept of neural reuse: the PB participates sector-wise to visual orientation and, as a whole, to mechanosensory-induced orientation. The PB is then involved concurrently (i.e. re-used) for different tasks.

Looking at other insect species, there are examples in literature where the role of the PB is not only limited to visual orientation with regard to landmarks and objects: locusts are an example where polarized-light based orientation needs an intact PB. Therefore the PB can be considered as a quite general centre involved in orientation behaviours where different sensory modalities can compete or co-operate with each other. Visual orientation requires a topographical map within the PB, whereas mechanosensory processing, in principle, does not. However, a touch of frontal body parts prompts the fly to escape backward, whereas a touch of rear body parts would trigger escape in the heading direction. Moreover, wind-sensing is direction-sensitive.

Whether or not mechanosensation makes use of a topographical representation in the PB, there is a possible need for integration with the visual sense within the PB glomeruli and mechanosensory orientation will need a complete PB structure, too. Admittedly, this issue requires further investigations through experiments. Another important aspect to be considered is related to the coordinate transformation between the map of mechanoreceptors on the body and the Cartesian direction in the world related to orientation and escape strategies.

After these considerations, experimental evidence in the fly suggests a possible involvement of the CX in a mechanosensory pre-processing focussed on orientation. In fact, whereas thoracic ganglia are responsible for eliciting a family of basic reflexes directly and locally to the leg system, on the other hand, signals, whose specific pathways are up to now unknown, reach the PB, which acts as a higher level orientation controller. This hypothesis is in line with the preliminary experiments presented, where flies show escape behaviours upon both touch and visual stimuli. The touch sensors in the leg can therefore elicit local stimuli in the thoracic ganglia (and so also in the leg-wing system), but also reach the PB in order to elicit orientation strategies normally used in visual orientation. This will become clearer after the following paragraph, which clarifies the role of mechanosensors in olfactory orientation.

These preliminary analysis can guide the formulation of a plausible model for a mechano-sensory orientation strategy.

6.3.2 Olfactory and Mechanosensory-Based Orientation

Thanks to olfactory orientation endowment, the fruit fly is able to track odour plumes under difficult wind conditions as well as under time-varying visual conditions. At the same time odour recognition takes place through a series of processing stages.

Recent studies [7] demonstrate that following an odour plume is not a simple task; at the same time also mechanosensory and visual inputs are required during flight. In fact, it has been shown that if an insect flies within a uniform visual panorama where high-contrast visual stimuli are absent, it is no longer able to orient toward the odour plume.

Multisensory integration is therefore an important ingredient of the fly behaviour as also shown in the model proposed in [11].

During the processing of olfactory signals, the olfactory receptor neurons (ORNs) transfer the acquired information to specific glomeruli in the AL (see Fig. 6.2). Moreover, projection neurons (PNs) connect the AL to both the LH and the MB calices. The LH is believed to receive also visual information from the ventrolateral protocerebrum (vlpr in Fig. 6.2): both visual and olfactory signals are delivered to the antennal mechanosensory and motor centre (AMMC) [7]. Antennal motor neurons arising from the AMMC, are also assumed to innervate muscles actuating the antenna.

On the other side, the passive motion of the antennae induced by wind stimulates neurons controlled by Johnston's organs (JO) situated in the area of the antenna cuticle which is deformed by air motion. This activates different areas of the AMMC. Interestingly, it is argued in [7] that the increased activation of the left (right) ORNs and PNs in response to an odour on the left (right) and via LH-AMMC neurons triggers an asymmetrical activation of the antennal motor neurons (AMNs) arising from the AMMC which, via antennal muscles actively rotate the antenna segments, mimicking a passive wind stimulus and using the same neural assemblies to trigger a leftward (rightward) AMMC mediated motion. This hypothesis, directly supported by experiments and by the neuroanatomical model of the olfactory orientation system, leads to include a direct interaction between mechanosensory-mediated orientation and the olfactory orientation. It is interesting to notice, how mechanosensors can trigger also orientation behaviours. On the basis of the models and experiments discussed, tactile information follows the same paths used by olfactory and visual stimuli to perform orientation strategies. The same neural structures are therefore shared and used for different behavioural tasks.

In this complex orientation task, also visual stimuli are required, as outlined above. As is well known, the higher processing of visual orientation is mainly handled by the CX that can guide the insect towards the most attractive object in the visual field. Flies are attracted by the nearest object located in the frontal part of the visual field, whereas an object seen on the periphery of the visual field ($[-140°, -170°][140°, 170°]$) triggers an escaping reaction. As already mentioned in the book, the fan-shaped body (FB) is the area where rewarding and/or punishing visual features are learned and memorized. The path followed during olfactory orientation starts from the olfactory receptor neurons that project to the antennal lobes. The LHs are responsible for the inborn, mostly reflexive behaviours triggered by olfactory signals, whereas the MBs are the centre where olfactory associations are created through reward- and punishment-based learning.

Olfactory-gradient orientation, in particular for a flying insect, as recalled above, is not obtained in absence of visual stimuli (i.e. when an uniform visual panorama is present). The LHs are also generating event-driven inhibitory input for the MB through extrinsic neuron connections (MB-EN2). Functional feedback connections from MB lobes centrifugally out to the AL via extrinsic MB neurons (EN-MB1) are

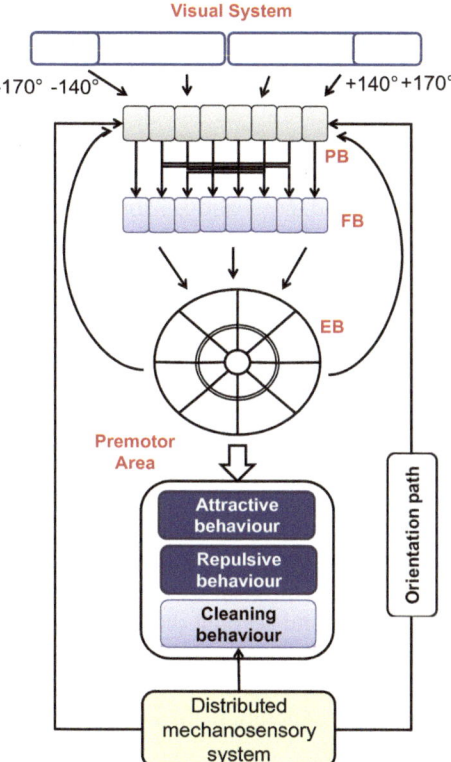

Fig. 6.3 Block scheme of the role of CX in visual and mechanosensory information processing. Visual information is handled by the protocerebral bridge (PB), the fan-shaped body (FB) and by the ellipsoid body (EB). The PB is mainly responsible for spatial identification of objects and targeting, in the FB the object features are stored and a meaning is associated to them using rewarding or punishing signals. The EB is responsible for spatial memory and in absence of fresh information, the object of interest can be tracked. The distributed tactile system can either locally trigger behavioural responses like cleaning behaviours or transfers the detected event characteristics to the PB, where orientation behaviours can be initiated as happens in presence of visual targets

also part of the system: a gain increase in the AL caused by this modulatory influence is at the basis of the formation of expectations.

For olfactory and mechanical stimuli acquired from the antennae, also the neural paths used for visual-based and tactile-based orientation are shared.

Looking at the CX in detail, in Fig. 6.3 a block-size model of the integration between the visual system and the mechanosensory system is depicted. From behavioural experiments and neurobiological analyses [10, 18], some roles of each subunit of the CX for visual learning and orientation have been identified. The fly's visual field, in the azimuth range $[-170^{\circ}, 170^{\circ}]$, is acquired by the compound eyes and relevant information are transferred from the PB to the FB for the generation of control signals sent to the ventral lobe for solving visual orientation tasks. The EB holds a short-term spatial memory. When the chosen target is temporarily obscured in the field, the EB projects a ghost copy of the missing object to persist in the following action for a while until new reliable information can be acquired by the sensory system.

In relation to mechano-sensors located on the insect body and legs, we can distinguish two main paths: a local handling and a transfer to higher brain centres, where orientation behaviours are elicited. Mechano-sensors are fundamental for the acquisition of the information needed during locomotion to coordinate and stabilize the walking gaits [5, 8]. In the presence of a tactile stimulus (e.g. an air shot) on a leg, the fly, depending on the intensity of the stimulus, either shows a local reflex like cleaning behaviour performed on the stimulated leg by an adjacent one or performs a repulsive orientation behaviour trying to avoid the source of the undesired contact. The orientation behaviour is obtained by transferring the local tactile information to the brain. From neurogenetics experiments the role of the PB in this process has been identified and this verifies the hypothesis that already existing neural paths can be reused for different sensory modalities, also taking into account that multiple stimuli can be integrated to take a decision.

In conclusion, the distributed sensory system and the neural structure devoted to handle this multimodal input are at the basis of the behavioural response in insects as well as in other animals. Taking into account the currently available information on the neuronal structure of *Drosophila melanogaster*, a plausible model of the dynamics involved in sensory processing mechanisms has been designed and discussed. The sensory modalities taken into consideration are those ones largely used by insects to survive in unstructured environments: olfactory, visual and tactile. The processing stages identified in the proposed model underline the basic principle according to which different sensory modalities are processed, at least in parts, within the same neural circuitries. In particular the orientation behaviours follow different neural paths: through the AMMC if the antennae are involved (e.g. olfactory and wind induced behaviour) and through the CX in presence of visual stimuli and tactile events detected on the insect legs.

6.4   Concluding Remarks

The examples reported in the previous sections of this chapter clearly show that in *Drosophila*, neural circuits, established for one task, are re-used and exploited for different tasks. In the animal kingdom, this transformation or adaptation, as Anderson introduced it, can be attained either during evolution, or during the development of the individual [1], without necessarily losing the original functionality of the neural circuits. This is exactly our case, where, not considering here which functional sensory-motor task has evolved before another during the early stage of the animal evolution or ontogeny, the different functions of the same neural structure are maintained to serve the specific behavioural needs. On the other side, it is well known in the mammalian brain that high-level cognitive functions require the cooperation and the concurrent activation of different areas of the brain typically devoted to other functions. This is of course a typical case for complex dynamical systems, where the single unit shows a non-linear dynamics which could be completely different from that one shown by the whole system. The massive re-deployment of neural structures is evident in higher brains, but it is difficult to be inspected in details. So, *Drosophila melanogaster* is an optimal candidate for this type of investigation, given its small brain and somewhat simple organization. But the smaller the brain, the larger is the need for reusing neural structures that are already working. Insect brains have a suitable size to investigate neural reuse in action. In the course of the chapters of our book, first preliminary models for addressing spatial memory, motor learning and body size were introduced, followed by a model in Chap. 5, that is able to cope with such tasks as motor learning, decision making, sequence and sub-sequence learning all-in-one. The common denominator in solving these diverse tasks is a unique core, a model of the structure and dynamics of the MBs, that has been considered as a specific multifunctional structure, that is at the basis of a number of different behavioural tasks that can be implemented concurrently. This is relevant, since MBs indeed are known to take a key role in many other functions than olfactory learning and memory (as seen in Chap. 1). The computational model presented in Chap. 5 was conceived to emphasize the role of MBs as the specific place where multi-modal associations are formed. The rich, spatio-temporal dynamics formed therein, potentially unstable, but tamed by the periodic inhibition induced by the LH, is exploited, via different readout maps (mimicking the MB output neurons), to serve a number of associations with other conditioning signals, coming from other parts of the brain. In that way, the same structure can serve at the same time different tasks concurrently. In our model we explicitly concentrated on decision making (classification), motor learning, sequence and subsequence learning, delayed matching to sample, attention, expectation, and others. Sequence learning could be further developed to implement an efficient capability of planning ahead, a key characteristics of cognitive systems. Our model represents the computational implementation of the concept of neural reuse in the insect brain: MBs are considered as the generators of a rich dynamics, controlled by sensory stimuli. Such dynamics is then shared and sub-divided, according to specific associations, and deployed for different concurrent tasks. We expect that the new view on the MBs will trigger inspection by neurobiological experiments and that the outcome will confirm this concept of re-use. Mutual benefit is expected from producing novel neurobiological details, and, at the same time, preparing more detailed simulations. The outcomes of the computational models could provide new ideas for the design of specific experiments focussed on revealing new impressive capabilities of the insect brain.

The model described in Chap. 5 takes only a small portion of the fly brain functions outlined in Chap. 1. For example, at the end of Chap. 1 it was underlined that, long-term autonomy will comprise additional interesting behaviours that will have to be fundamental components of the adaptive robot of the near future. One of these behaviours is novelty choice, a key-component of curiosity. This was recently addressed as a result of the interplay found between CX and MBs [16]. Currently, efforts are made to assess novel results from the biological side and to transfer them into efficient models which could contribute to build efficient autonomous machines.

In the eye of the authors, the insect brain represents one of the best places to identify the basic rules and find the solution to build a completely autonomous machine. The treasure of behavioural capabilities, so far identified in insects, is expected to further increase in the near future. Even complex behaviours that, judged from today's view are unlikely to be shown by insects, could be found, with suitable experimental design. One example is the discovery of the capability to solve the Morris water maze task in cockroaches, crickets and *Drosophila*, a task involving complex spatial memory. Insects really are the model organisms of choice, able to inspire the design and realization of the most advanced all-terrain robots of the near future. This will require efforts from a multi-disciplinary perspective in order to achieve a solid advancement of science and technology.

References

1. Anderson, M.: Neural reuse: a fundamental organizational principle of the brain. Behav. Brain Sci. **33**, 245–313 (2010)
2. Arena, P., Patanè, L.: Spatial Temporal Patterns For Action-oriented Perception in Roving Robots II. Springer (2014)
3. Arena, P., Patanè, L., Termini, P.: Learning expectation in insects: a recurrent spiking neural model for spatio-temporal representation. Neural Netw. **32**, 35–45 (2012)
4. Aso, Y., Hattori, D., Yu, Y., Johnston, R.M., Iyer, N.A., Ngo, T.T., Dionne, H., Abbott, L., Axel, R., Tanimoto, H., Rubin, G.M.: The neuronal architecture of the mushroom body provides a logic for associative learning. ELife **3**, e04577 (2014)
5. Büschges, A., Gruhn, M.: Mechanosensory feedback in walking: from joint control to loco-motor patterns. Adv. Insect Physiol. **34**, 193–230 (2008)
6. Dean, J., Kindermann, T., Schmitz, J., Schumm, M., Cruse, H.: Control of walking in stick insect: from behavior and physiology to modeling. Auton. Robot. **7**, 271–288 (1999)
7. Duistermars, B., Frye, M.: Multisensory integration for odor tracking by flying *Drosophila*. Commun. Integr. Biol. **3**(1), 60–63 (2010)
8. Dürr, V., Schmitz, J., Cruse, H.: Behavior-based modelling of hexapod locomotion: linking biology and technical application. Arthropod Struct. Dev. **33**, 237–250 (2004)
9. Frye, M.: Multisensory systems integration for high-performance motor control in flies. Curr. Opin. Neurobiol. **20**(3), 347–352 (2010)
10. Hanesch, U., Fischbach, K-F., Heisenberg, M.: Neuronal architecture of the central complex in *Drosophila melanogaster*. Cell Tissue Res. **257**, 343–366 (1989)
11. Ito, K., Suzuki, K., Estes, P., et al.: The organization of extrinsic neurons and their functional roles of the mushroom bodies in *Drosophila melanogaster* meigen. Learn. Mem. **5**, 52–77 (1998)
12. Liu, G., Seiler, H., Wen, A., Zars, T., Ito, K., Wolf, R., Heisenberg, M., Liu, L.: Distinct memory traces for two visual features in the *Drosophila* brain. Nature **439**, 551–556 (2006)
13. Nowotny, T., Rabinovich, M., Huerta, R., Abarbanel, H.: Decoding temporal information through slow lateral excitation in the olfactory system of insects. J. Comput. Neurosci. **15**, 271–281 (2003)
14. Ries, A.S., Hermanns, T., Poeck, B., Strauss, R.: Serotonin modulates a depression-like state in *Drosophila* responsive to lithium treatment. Nat. Commun. **10**, 15738 (2017)
15. Scherer, S., Stocker, R., Gerber, B.: Olfactory learning in individually assayed *Drosophila* larvae. Learn. Mem. **10**, 217–225 (2003)

16. Solanki, N., Wolf, R., Heisenberg, M.: Central complex and mushroom bodies mediate novelty choice behavior in *Drosophila*. J. Neurogenet. **29**(1), 30–37 (2015)
17. Strausfeld, NJ, H.F.: Deep homology of arthropod central complex and vertebrate basal ganglia. Science **340**, 157–161 (2013)
18. Strauss, R.: The central complex and the genetic dissection of locomotor behaviour. Curr. Opin. Neurobiol. **12**, 633–638 (2002)
19. Zill, S.N., Moran, D.T.: The exoskeleton and insect proprioception. III. Activity of tibial campaniform sensilla during walking in the American cockroach, *Periplaneta Americana*. J. Exp. Biol. **94**, 57–75 (1981)